It's All B. S.

(Biological Stuff)

Gretchen Watkins

IT'S ALL B.S. | (BIOLOGICAL STUFF)

978-1-955595-00-1 (Hardcover)

978-1-955595-01-8 (Paperback)

FIRST EDITION

Table of Contents

1

Scope and Sequence for Understanding Biology Stuff

The Adventure of Understanding Life

Every person in the United States is now expected to graduate from high school in order for them to understand a certain amount about life science. It has taken me a lifetime to understand as much as I do now, and I am still learning. Surely, it has been an adventure to understand life. I am a biology teacher myself. Biology is the study of life, so it is my profession to understand it. But man, it really is complicated! The new standards developed spell out what each student should be proficient in, when looking at information related to life science. I just completed my fourth year as a science teacher, and I feel like I too graduated along with the freshmen I have taught through seniors. I fully know my science content that a high school graduate should know. I have come to this place of writing so others may also have this knowledge as they navigate their own lives. It is not just the biology content but the true examination of life that I hope will help during the struggles of life. By the end of reading this book you too will understand what a new high school graduate should understand about life science.

Introductions first

I am actually a scientist and engineer, by training and by natural tendency. I love to understand how things work, and figure out new ways to make them better. I am relatively new to being a teacher. I worked professionally as an engineer/scientist for around 18 years prior to my high school teaching position. I worked on behalf of the federal, state, and tribal governments to keep our natural environment as healthy as possible for living beings. Through these jobs, I found that many people really do not understand what it takes to be healthy and happy. Politics and other agendas take priority over making decisions for the healthiest lifestyle. This was very hard for me to accept as an engineer, as I was naive and thought if the numbers supported health, then all subsequent decisions would be made based on the best balance of resources. As a scientist, I understood that science is about balance, or homeostasis as we biologists like to call it. Balance is the key: not too much sugar, but you have to have some in order to have energy. As an engineer, one is trained to find balance with numbers. The sweet spot where everything works out just right. Well, that is all B. S. when you really think about it. How can two adjoining cells maintain balance when they are both in need? Yet we have muscle tissue that works in harmony with our other body systems as it together powers our movements. It is the glory of Biological Stuff.

The legend of the title

My husband came up with this title, "It's all B. S. (Biological Stuff)". I have been married almost 18 years. We met in graduate school; he was getting his Masters in Forestry and Landscape Ecology, and I was getting mine in Environmental Engineering. The difference in the disciplines is that while he was learning about the ecological system, I was applying our understanding to control the system. To this day, we differ in our overall take on the world. I want specific solutions and he wants to understand the world. Understanding the big picture is all fine and well, but the application of the understanding is what I love. I want control. So when he told me his title of this book *he* was going to write, I had to take it to the next level. As all good relationships go, when you steal your husband's book title and write it in your own way, there is a bit disappointment awaiting from your husband's side. But I say this now: "Thank you husband for the inspirational title, and for our constant dialog about how ecological systems work." My husband, Radley, is a big-picture thinker, seeking knowledge on how complex organisms interact to survive. I need to break it down to the most fundamental level, to understand the complexity of life. Rad keeps having me learn the big picture, and I apply it to maintain balance within my life and balance with those that I interact with so that I can find harmony in life.

Understanding life just helps us apply the concepts to our unique struggle. This book will start off by teaching you all about chemistry. You might stop reading once you learn how biochemistry works to keep you healthy and strong-minded. But it is a bit more complicated than understanding chemistry, or else we would not eat fast food.

Then we will move into how the chemistry comes together to make structures in our bodies that provide certain critical functions. These functions are one's specific characteristics and traits that make them who they are. White, black, green, strong, and smart are all traits that are just made up from chemicals. These characteristics come together to make the organism whose sole purpose is to have the energy to live. Energy is chapter 4, and one of my favorite units to teach. Without energy how can we love, propagate, play music, or perform any of the many other gifts of life?

This will then bring us to communication, because we must work together to save energy or build energy. Nerve cells must communicate with muscle cells to make you work, just as bankers must communicate with real-estate agents to buy a house. Once we understand the *purpose* of life we will investigate the *how*. How do these chemicals like DNA make the characteristics, how are they regulated, and ultimately how are they passed down? Propagation and evolution over time, create the array of diversity our world has to offer, and how the various organisms interact. Lastly, we will discuss natural selection and how ultimately life is about community relationships.

What you will learn in this book is the basics of life that we try and teach every American, so they can grow up and make decisions that are good for their own health and for those around them. This is not the biology that older (pre-2000 high school graduate) learned. Pre-2000 biology was about categorizing living things into similar structures and memorizing the structures. Now life sciences is based in molecular biology as we have learned the role of DNA and RNA. This biology will give you a base for understating how life works so you can critically think about life as a new understating of biology. The medical field is constantly changing how we understand life so we need to be critical thinkers to make good decisions. For example, communicating with a doctor or looking at lab results as you or a loved one gets sick is always challenging, but if you do not know how life works you cannot make the best decisions or ask the right questions as you go through the process. This book will help you have a good life science base when you are deciding all sorts of things; like what groceries to buy or voting for a representative. Life is a series of choices that we get to make. It is up to us to make informed decisions for the best life to live so that we can figure out things based on the unique organism we are. This book will help give you the biological basis for good decisions for your life.

As you reflect on this chapter, ask yourself what is the meaning of life? Why is it important to understand how life works? If you are a parent or guardian of a student in school taking biology or advanced placement biology, use these questions as discussion starters. I find going for a walk or prepping a meal with my teenager is the best time to have focused conversations. True learning happens when you can talk through the thoughts that come up as you learn and process new information.

2

It's really all Chemistry

Chemistry of life is broken down in to 4 types of molecules that are important. We generally know 3 of them as **fat**, **sugar**, and **protein**. They are just the things we eat. The other one is **nucleic acids** or we might know as the building blocks of DNA and RNA. When we talk about chemistry, some of us think of nutrition, some drugs, and some the feelings we get when in love. All of these categories actually all work on the same premises of chemicals/elements. Biologically, the most common elements are **carbon**, **hydrogen**, **nitrogen**, and **oxygen**.

I was raised a protestant, confirmed as a teenager and all the normal church stuff. It gave me some comfort that if I did die then I would go to heaven because I was a good kid. Then one day after church, when I went out to shake hands with the minister, I gathered up the courage to ask, "Is there really a heaven?"; in the sermons, there was not much mention about heaven, so I felt I had to clarify. The minister stood there holding my handshake and looked me in the eye and said, "I don't know." Maybe it was the best gift I was ever given, as it sent me on my journey of wonder.

The period of periodic table

High school chemistry was right around the corner in my life. My father is a physicist, so I had learned a fair amount about chemistry. My dad even has the whole periodic table of the elements memorized. I never did accomplish this, a great disappointment in my life. One day I will do it. So I had quite a "golden backpack" going into high school chemistry, which was good because I had really struggled in school up to this point. I mean I was good in math but that was it. Now the tables were turned. All those kids that had been quite good at school, they now did not understand chemistry. To me it was just math applied to the amazing system of the periodic table; elements that are organized based on how they will react. Nothing to memorize as long as you understood the system on the table.

In chemistry class, we could take our understanding of the periodic table and create reactions and predict correctly what would happen. This fit well with my dislike for uncertainty. It amazed me even more to know that all of the world, even the universe was made up of these elements that I understood. I could mathematically predict the outcome and get darn close if I was meticulous about my lab technique. I was still freaked out about dying but at least something made sense in the world.

The periodic table only has a few big players in the game of life. As mentioned before, we have carbon, hydrogen, nitrogen, and oxygen that are biologically important. Everyone knows life is made from Carbon. But what is it about carbon that makes it so useful? Hydrogen and oxygen make up water (the key to life) if in a two-to-one ratio, but if in a two-to-two ratio, it is hydrogen per oxide, and it makes cells die. If we look at where these elements are on the periodic table we can start to tell the story.

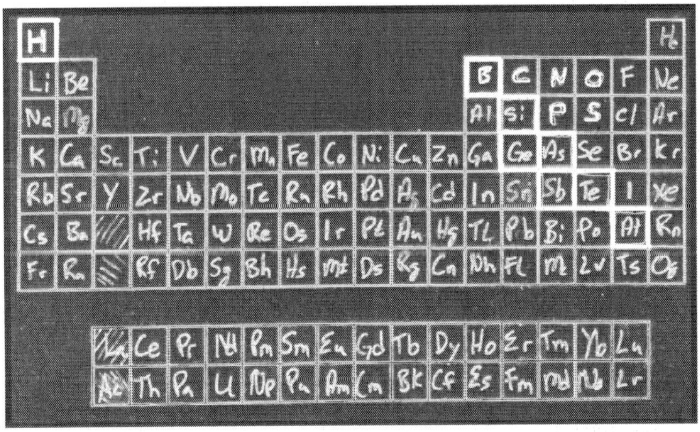

Figure 1 Periodic Table

The life of and from Carbon

Looking at the periodic table you can see that all of the biologically important elements are clustered in the upper right-hand corner. Carbon is on the left side of the dark stair-step line that divides the metals from the nonmetals. Metals, or as chemist, call them **cations,** lose their electrons when bonding, and non-metals or **anions** gain an electron when bonding. The farther you are away from the stair step line the stronger the element is at stealing the electrons. Carbon being right on the line is the happy 'get-along' element that shares electrons instead of stealing them. Carbon is not too dominant, making it able to bond with many elements, including the super biologically important ones, and even with itself. Carbon is such a "get-along" element, that it both forms long **chains of carbon** and can also bond in other elements. Carbon having 4 valences (outside) electrons gives it a means to share 4 electrons, giving it diversity in bonding ability.

My name is Bond (enter all the types of bonds)

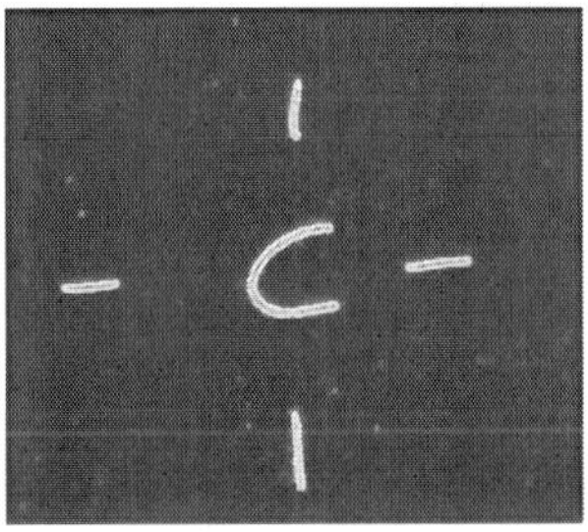

Figure 2 Carbon

Before we move on, let's take a step back and review what an element is and how bonding works. Elements are the simplest form of matter, which cannot be broken down any further by chemical means. You can bust them up with a nuclear reaction, where the atoms nucleus, which is made of neutrons and protons, comes apart. The atom is defined based on the number of protons (Atomic Number) in the nucleus. One proton and you have Hydrogen, two protons and you have Helium, and so on. The nuclear reaction in sun entails combining the nuclei of Hydrogen to make Helium by nuclear fusion. But I digress, back to chemical reactions or bonding. The atom's nucleus has protons that have a positive charge and neutrons with no charge. Then flying around the nucleus is the electron that has a negative charge. The protons and electrons tend to balance so the make the atom neutral. Here is the key to bonding; the protons have to stay put or it will become a different element, but the electrons can move about. So these electrons are attracted to the protons in the nucleus, but the electrons on the outside (valence electrons) can be attracted by another atoms nucleus to the point that sometimes they completely transfer to the other atom leaving the original atom with a positive over all charge (more protons then electrons). The new atom that gained the electron then attains an overall negative charge. The two atoms are then associated by their charges like Sodium Chloride (salt). This stealing of electrons is called ionic bonding. Most living matter uses another kind of bonding called covalent bonding. It is very similar but instead of stealing the electron the two atoms share the electron. Only non-metals can share and form covalent bonds. Chemical bonds can be broken and changed, and in doing so energy is exchanged and the electrons are exchanged. Carbon is the best at making covalent bonds and thus why it can make the long chains of carbon.

Fats – the inside story

Fat is the simplest form of carbon chain. It is just carbon chains with hydrogen filling in extra bonding areas. Living things use fat to store energy. The energy between the bonds is called potential chemical energy. **Potential energy** is just another term for stored energy. The best example of high potential energy is a ball on the top of a hill. The active energy, called kinetic energy, is the ball rolling down the hill or the muscles contracting to run.

Fat is how we store energy in the carbon bonds. There are different types of fat. **Saturated fats** are carbon chains that are saturated with hydrogen. There is only a single bond between the carbons forming straight chains while hydrogen are bonded to the remaining locations. These saturated fats then line up more easily and are often solid at room temperature. Butter is my favorite example of saturated fat (and the best tasting). **Unsaturated fats** then are often liquid because they have some double bonds and less hydrogen bonded, making them more fluid. Unsaturated fats are typically healthier than saturated fats. Humans wanted healthier fat but wanted it solid so they could spread it on bread. So they added the hydrogen into the unsaturated fats making them solid. Adding the hydrogen in the carbon chain is call hydrogenation and the products are thus called partially hydrogenated oils. These hydrogenated oils, however, ended up being the unhealthiest of all the fats. The general rule of thumb for food is, the less processed by humans the better.

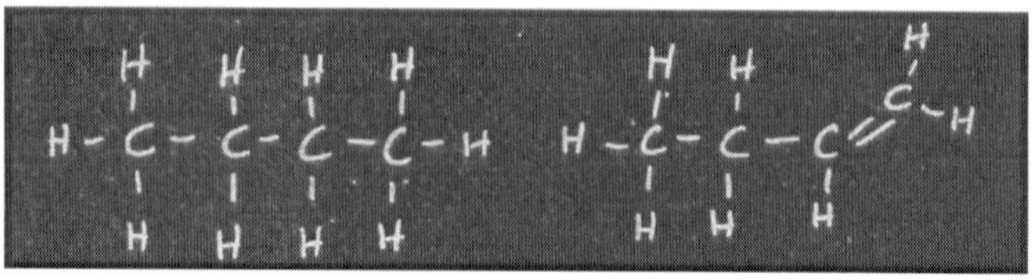

Figure 3 Saturated and unsaturated fat

When I graduated from Rensselaer Polytechnic Institute (RPI), I interviewed for a company that wanted me to make a fat that did not make people gain weight. They paid well and the biochemistry was interesting, but I just could not take the job. Instead I went on to get my Master's degree in environmental engineering, working on supplying New York City population with clean drinking water. Working in the environmental world helped me learn the larger systems and ecology, but my foundation of molecular biology always served me well. If you know how the small stuff works, you can always figure out the big stuff.

Sugar (not spice), and everything nice

I, like most people, worry about eating healthy. Most of us understand that fat is a calorie rich food (note: calorie is a unit of energy) and we try to minimize the amount of fat that we consume. But life is about balance and our systems have evolved to crave calories, as it motivates us to hunt and gather, farm, plan, and store, so that we do not starve. Now, in this modern food economy, getting our calories requires little energy-output-to-calorie ratio, and that means we have to moderate ourselves and find the proper balance.

The other calorie-rich resource is sugar. All life starts with sugar created from plants by converting sunlight energy into carbon bonds. Plants utilize the sugar that they make for energy, to grow and reproduce. The plants will also convert the simple sugar into starch or vegetable oil to store for later growth, or they convert it into protein for the plant's living processes. Living things that cannot **photosynthesize** must consume the plants for their sugar, oil, and protein.

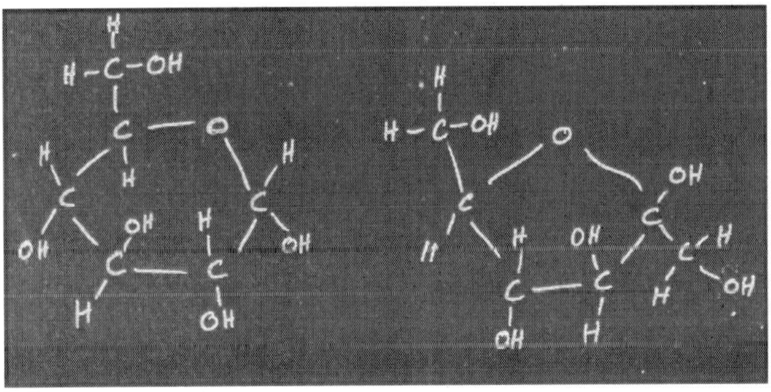

Figure 4 Fructose and Glucose

Photosynthesis is the process of taking a single carbon, bound to two oxygen molecules (Carbon dioxide, CO_2), and using energy from the sun to bond the carbons together. This chemically stored energy between the carbon bonds is then used by all living things to survive.

In my house, my husband is the sweet tooth, and I am the fat tooth. When I am hungry I crave butter, cheese, and fried things, or better yet fried cheese. (I have yet to have a fried butter stick; I think that is just taking it too far. I have to maintain some self-control now.) My husband on the other hand will eat cookies and cake. Both items have some sugar and some fat, just in different proportions. His food has more sugar and mine has more fat. Sugar and fat go well together.

Cells are the smallest unit of life. All cells (plant, animal, and even amoeba) are made up of part sugar, fat, and protein. The smallest of cells are bacteria. There are about a trillion species of bacteria in the world, each one specializing in a different way, some good for us, and some bad. All have a **fatty membrane** that keeps the inside on the inside and the outside on the outside of the cell. All cells have this fatty membrane and it is one of the reasons we need fat in our diet. Other reasons are the need for insulation and stored energy. But this membrane helps the cell take in what it needs while getting rid of waste.

Protein – it's in the little things

Cells use sugar for energy; some cells (photosynthetic) create the sugar they need, and others (consumers) "eat" the sugar they need. Each cell uses the sugar for energy, which then drives protein to do the work of the cell. Each cell specializes in some way to create what it needs and maintain the environment on the inside of their cell that works best for them. Protein is *that* worker in the cell.

Muscle cells are a great example of cells that do work. We know that when we want to build muscles we need to eat protein. Chicken breast (the large chest muscle used to flap wings) is a great source of protein. And we know also that when we eat a lot of sugar we get all hyper and love to run around - well at least when we were young anyway. On the cellular level, proteins in the muscles are burning energy (sugar) to make the movement.

Protein is made out of **amino acids.** There are 22 of them that are the building blocks that make up all of the protein we have in the living world. Amino acids are also a carbon based molecule, but in addition to carbon, amino acids also have nitrogen and some of them have sulfur too. Amino actually means nitrogen. Nitrogen sits one to the right from carbon on the periodic table. The rules of the periodic table tell us that as we move farther to the right from the stair step line, the better the ability is for the element to attract the electron. Nitrogen is still close to the transition line so will share electrons instead of stealing them. Thus nitrogen line carbon is a good building block element, but it also attracts the electron more than carbon. The other part of the amino acid is the acid part. Acid is a chemical that donates a proton (hydrogen) or accepts an electron. To understand this concept we must now look at hydrogen on the periodic table.

Hydrogen is way up on the top left of the periodic table. It is the first element. The elements are arranged in how they react to each other. And they are numbered based on how many **protons** they have. Hydrogen, being the first element then has one proton. Protons are positive and are in the center, or in the nucleus, of the elements. Electrons are negative and fly around the outside of the element; as the negative is attracted to the positive it holds together in that fashion. Hydrogen with just one proton can easily lose its electron. So when we talk about **acids** we are normally talking about hydrogen, and once they lose their electron, they are then just a proton. You might think of something that is acidic as having a low pH; the H being Hydrogen. Therefore, pH is the measure of hydrogen ions within a solution.

Lastly, when we look at the amino acids, we should note the oxygen. Oxygen is the next one to the right from nitrogen and has a strong attraction for electrons. Oxygen attracts the electron away from the hydrogen giving the oxygen an overall negative charge and leaving the hydrogen as a mere proton. We will get more into this when we talk about energy but right now, I just want you to recognize that this set-up of the different elements in the amino acids gives each one a different property. Some have a negative charge and some have no charge at all.

These different properties are oriented in such a way that when you put the amino acids together to make proteins you have different functions of the proteins, depending on the order of the amino acids. Proteins are the workers of the cell. So depending on how we put the amino acids together, we can have proteins that make muscles move and make DNA.

When you look at what a cell is made of other than water, it is sugar, fat, and protein. Depending on the type of cell, you will have these in different proportions. Each of these **macro molecules** is needed for life to exist. None of them are bad; they each have a function. Depending on your own activity and body chemistry, you will metabolize these molecules in different ways. In general, you should think of consuming about a third of your calories from each of these macro molecules. The other part of the cell that we have not discussed yet is water.

Water - you doing today?

Water is oxygen and hydrogen. About 80 percent of a cell's content is water, depending on the type of cell. Water is what all of the macro molecules move around in. Sugar is soluble (dissolves) in water and fat is not. Some proteins are soluble and some are not. Some proteins move water from one side of the cell to the other side, but almost always, water is critical to life process. That is why we look for it in outer space if we are looking for life. We mentioned how important the movement of electrons is between the elements. The movement of electrons between the hydrogen and oxygen is what gives water the properties that sustain life. Oxygen attracts the **electrons,** giving the oxygen a partial negative charge, and the hydrogen provide a partial positive charge. Having one side positive and one side negative makes the molecule polar and gives it the ability to attract other charged particles. This attraction between the charges is what holds water drops together or helps to dissolve other charged molecules.

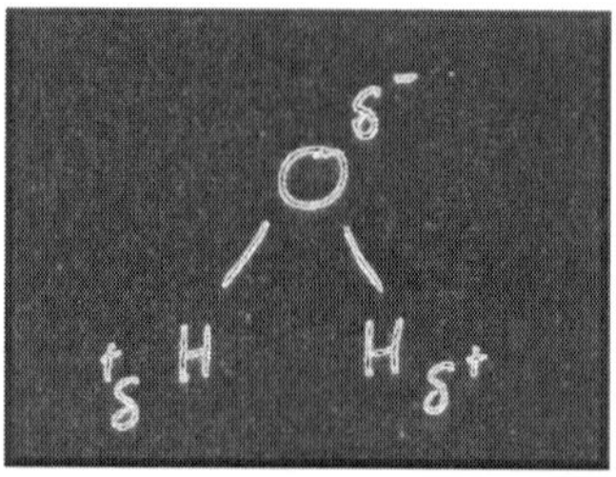

Figure 5 Water Polar

The attraction to the partially positive hydrogen is called a **hydrogen bond**. The hydrogen bond is important for water but it is also important for the last of the biological macro molecules, DNA or deoxyribonucleic acid. There are 4 monomers (building blocks) for DNA namely Adenine, Thymine, Cytosine, and Guanine -- A, T, C, and

G. Adenine, hydrogen bonds to Thymine and Cytosine hydrogen bonds to Guanine. This is the code of life. Three of these letters together will code for one of the amino acids. The DNA code tells what order the amino acids must be in so that the cell can make the proteins that it needs. This is such an interesting process we have an entire chapter later talking about protein synthesis and DNA. Our understanding of this process is just derived from the chemicals interacting --and now we have the power to manipulate the DNA to create an everyday fungus to a glowing pig.

The importance of chemistry in our understanding of biology is fundamental to understanding how biology works. My dad the physicist always said, "Biology is actually chemistry, and chemistry is actually physics, and physics is actually math". Definitely my understanding of chemistry has made me a much better biologist. Understanding the basics of chemistry, as described in this chapter, sheds light on how life works. These few elements hydrogen, carbon, nitrogen, oxygen, and sulfur -- are things that make life possible. Each element has unique predictable properties that when combined make molecules that are also unique and predictable. Our understanding of these macromolecules - fat (lipids), sugar, amino acids (protein), and nucleic acids (DNA) - tells us how life works and how we can manipulate them to find balance and harmony in life.

Good conversations starters for chemistry are to ask, how do elements interact to make molecules important to life? What are the 4 macro molecules and what's important about them, which is your favorite to eat? If plants don't eat fat where does vegetable oil come from?

3

Structure and Function

Old school biologists, loved structure. Structure is how something looks, like German Shepherd dogs have pointy ears that stand up. But the old school scientist often left out **function** (how it works), like what the pointy ears do for the dog. Also we did not have good microscopes until the 1900's, and research on the chemistry of life (molecular biology) was really just getting going in the 1950's. The **human genome project**, where we figured out the DNA of humans, was done in the 1990's. And now CRISPER, (clustered regularly interspaced short palindromic repeats) the gene editing tool, is just getting affordable enough for scientists to use it. Understanding of structure and function from the molecular level all the way up to the ecosystem level is important for us to truly understand how life works. In this chapter, we will look at the structure and function of cells, the smallest unit of life, and the basis of living.

Taxonomy – the system of clarification

Carl Linnaeus was a biologist in 1700's. He is known for creating the *system of clarification* based on the structures of living things. His system started with the two living categories: plants and animals. It worked by starting with structures that would make things unique from other things, and as you worked down his system, you could finally get to a completely unique species. The system started with kingdom that had, as mentioned, two categories -- plants and animals -- and ended with the most specific species. This system included all categorized life known at the time. Each taxonomic level became more specific with time. The system has been refined over the years to include new discoveries like bacteria and fungus that are not plants.

Taxonomic hierarchy (example: gray wolf or Canis lupus)

Domain (Eukaya), Kingdom (Animalia), Phylum (Chordata), Class (Mammalia), Order (Carnivora), Family (Canidae), Genus (Canis), and Species (lupus)

Students love **taxonomy**. They love all the different types of animals and figuring out what is different about them. My son is 11 years old, and we live on an island in the summer. He has grown up on the island his whole life. I went into labor with him on the island and my husband boated me off to the hospital. We all love the island, but my son's brain does not work like mine. I see systems and chemistry and he can see a green frog amidst an array of vegetation from what seems a mile away. When he was 3 years old he would slog off through the muck and vegetation as I sat in the canoe declaring I did not think there was a frog over where he was heading. He would find the frog and show me, to my amazement. It was not until he was 5 that I came to realize how different our brains were, when I found out he was color blind, which gives him a unique advantage in seeing camouflaged things. Our brains are different in many ways, I learned over time. He could read quickly and easily, whereas I struggled throughout grade school. We gave him all the identification books and he would study them before he even learned to read. Now that he is a superb reader, he knows all the names and habitat each animal inhabits. Sibley Guide to Birds is his favorite book, and he takes it with him whenever he goes into the wild.

There is something nice about organization. I categorize my clothes by having socks in one drawer and t-shirts in another. So, when I get up in the morning to go for a run, I can get dressed without really even thinking. Organization helps us to be efficient but also to start to understand the connections between items. What makes a bird different from a lizard? Most students should be proficient at categorization by the time they make it to high school biology. Yet taxonomy is a great place to start for the study of life. Asking the question "what is different and unique about this organism?" Helps us to move on to think critically, "Why are they different?" or "what role does this organism play?"

In the last chapter on chemistry we classified the different macro molecules that make up living cells. Hopefully when you read that chapter it got to you to questions like, "If I eat different proportions of sugar or fat, what will be the healthiest for my metabolism?" Science is the practice of asking questions. Classification is a tool to help us formulate our questions. Our brains are actually geared to classify things so we can simplify live to make decisions.

Bacteria and viruses – the common confusion

One question I want students to ask is, "With the amazing amount of diversity of life, how are all living cells made from the same basic parts?" All cells, from the simplest to the most complicated, have a few basic similarities. All cells have a membrane of fat, code of DNA, protein that enables work, and RNA or ribosomes that convert the code to the protein. **Prokaryotic** cells are the most simple of cells, and we know them as **bacteria**. Bacteria are the smallest form of life. We did not know about it until relatively recently. People still confuse them with **viruses**. The big difference between a virus and bacteria is that bacteria can live and replicate on its own, whereas a virus needs to go into a cell and use its ribosomes to read its viral code in order to replicate new viruses. Bacteria are alive because they have the parts to replicate independently; viruses cannot. Life is defined as the ability to replicate itself. So life is membrane, DNA, protein, and ribosomes, because only with all these parts does life continue.

If we start at the outside and work our way into the cell to understand it, we start with the **membrane**. The membrane is the part of the cell that keeps the outside on the outside and the inside on the inside. The way this works is based on the differences between the chemical properties of fat and water. Think of oil and water when you put them together, like in oil and vinegar salad dressing (vinegar is basically water). Oil and water are separate; you have to shake the dressing before you pour it on your salad. Water is polar with a positive side and negative side, and oil has no charge. Charged stuff likes charged stuff (the scientific term is hydrophilic) and non-charged stuff likes non-charged stuff (the scientific term is hydrophobic). Hydro means water and philic means love of; therefore hydrophilic means love the water and hydrophobic means scared of water. In reality, it is not much about love or fear but the simple attraction of positive and negative to the polar parts of water.

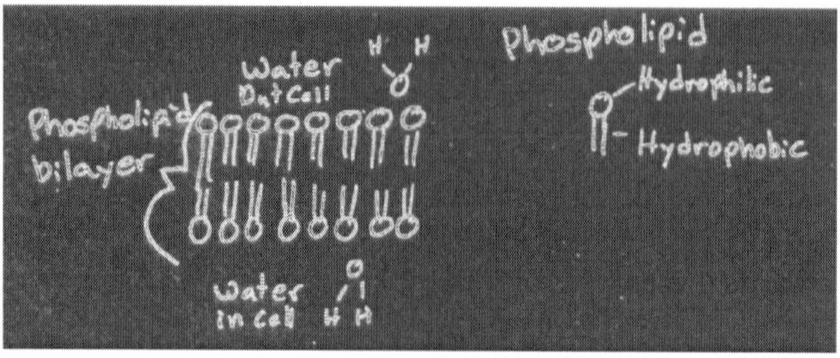

Figure 6 Water and Phospholipid Membrane

A dive into the terminologies

The membrane works based on the principles of hydrophobic and hydrophilic natures. The fatty membrane has a molecule called a phospholipid. The **lipid** is the fat (oil) and the **phospho** is a charged phosphorus and oxygen part of the molecule. Most of the molecule is 2 long chains of fat with a top charged part. The membrane is then a line of these molecules on one side, with lipid tails facing in towards another phospholipid, so the lipid tails are together and the phospho is towards the inside and outside of the cell. This phenomenon is called the phospholipid bilayer.

I know all this terminology gets very complicated; be it hydrophobic, hydrophilic, or phospholipid bilayer. But I am a nerd and I sit around thinking about these words and how they describe the workings of life. I find it very interesting that when I am given a problem, I can think of a solution to it with my understanding of life.

My husband, Rad, studied Forestry, as he saw it as the home for all the wildlife he loved. He has worked on all sorts of wildlife projects protecting everything, from marbled murrelets to grizzly bears. When we were dating, he lived on his family's island and he would bathe in the lake often. Even though he used camping soap he was a bit concerned about not being good for the lake. So he asked me what I thought: Is it safe to use soap in the lake if the soap is biodegradable? "Well," I started, "Soap is a *surfactant* meant to break down oils. Oil is an important part of cells, it holds them together with the *phospholipid bilayer*. The soap breaks down this protective membrane in sensitive organism like frogs that already have the most minimal protective skin, as they use their skin to breathe by diffusing oxygen through it. The soap could break this skin down and make the frogs even more susceptible to environmental contaminants. He was hooked at phospholipid bilayer. We use essential oils now for our kids to bathe in the lake.

Sorry for the side note, and back to the fatty membrane or phospholipid bilayer, and why it is so important to understand, as it is the way things go in and out of the cell. *Bad stuff out, good stuff in.* Small uncharged stuff can just diffuse in and out following the concentration gradient, meaning oxygen will move from a high concentration to a low concentration across the membrane. Thus as we use up oxygen, it will be low inside the cell so it will diffuse into the cell as we breathe and bring in new oxygen rich air. Bigger stuff or charged stuff cannot pass the lipid part of the membrane. Remember oil and water don't mix, or hydrophilic (charged stuff) and hydrophobic (oil, uncharged stuff) don't mix. So here is the cool part about life. There are proteins in the membrane that control stuff that goes in and out, like doors to your house. The protein is held in the membrane and can passively (no energy) let hydrophilic stuff in, following the concentration gradient or activity (using energy) the protein can push or pull the big or charged stuff into or out of the cell. Everything in life ends up revolving around this process of good stuff in, and bad stuff out. So, yes… life is a struggle, requiring constant maintenance of a favorable environment within the cell called homeostasis. Think of the cell as a house. We have a door to let the food in and a toilet to remove waste. If the toilet gets stopped up, soon there will not be a favorable environment within the house as the waste would build up.

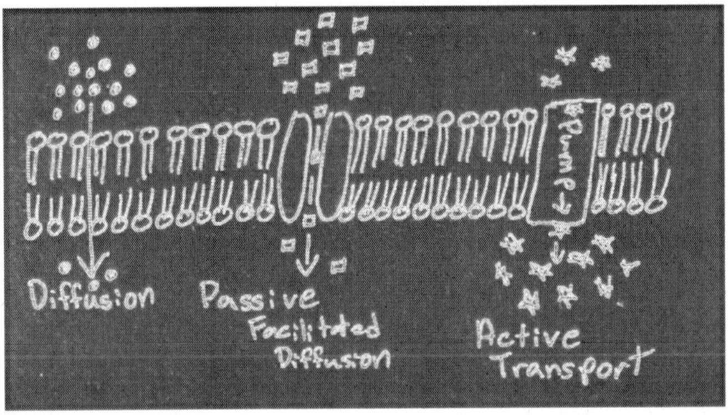

Figure 7 Diffusion, active and passive transport

Another analogy of the cell and a house is maintenance of a certain temperature. In my house we like the thermostat at 70 degrees Fahrenheit. If it gets too hot, the air-conditioner will turn on; if it gets too cold the heater turns on. Our bodies work in a similar fashion; if we get cold, we shiver to make heat in our body, and if we get hot we sweat to bring down our temperature. Our cells do this maintenance of a favorable condition by having proteins specialized for moving things in and out of the cell. One of the most interesting and well understood systems to maintain homeostasis is for sugar. If we eat too much sugar we run around like a crazy maniac, if we do not have enough sugar we get sad and lethargic. We will talk more about this system in the energy chapter. For now know that homeostasis is the maintenance of a favorable environment and the cell has certain types of proteins that will move all sorts of chemicals in and out, depending on what it needs.

Osmosis is the process of water moving from one side of the cell membrane to the other. Osmosis happens passively so that the water moves from high concentration to low concentration. Think of bacteria that are moving around our world, they occupy super dry areas like a door knob and wet areas, like your mouth. How can bacteria live in such varied environments? If water can just move from the wet outside to the inside of a cell freely then the cell would explode. Or vice versa, if a cell was in a dry environment it would just shrivel up. But cells are able to maintain homeostasis by changing the type of proteins in their membranes. If the cell is in a wet environment, it can make proteins that pump water out of the cell so it does not explode.

Responding to the environment is a key biological function. In complex organisms like our bodies, we have systems like our brain and nervous system to help us respond. When we grab a pot on the stove that is too hot, we let it go. All life has this ability of response to the environment. An example would be the bacteria that form proteins that pump water out when they are in wet environments. Changing the type and amount of proteins is the first way all life sets itself up to respond to the environment. A more complex example is the development of the nervous system. As the cells are growing in an embryo, some cells are going to differentiate to become nerve cells. It's like the organism knows it will need to respond to the environment so it sets up these systems to do it for them. The specialized cells like nerve cells just have a different set of proteins that form the cell so that the membrane can communicate with the membrane of an adjacent nerve cell to send a message like "the stove is hot."

Prokaryotic cells (bacteria) are simple. They are made of basically a membrane, proteins, RNA, and DNA. The RNA reads the DNA so the RNA can then synthesize or make the structure of the proteins, then the proteins go to the membrane to maintain homeostasis. Complex cells need a little more organization to them so that they have organelles made of membrane. Organelles are compartments inside the cell that have a special function, like the mitochondria (power house of the cell). The organelle that acts as the control center is the nucleus, which is a membrane that encloses the DNA, and provides protection to the DNA. This membrane that encloses the DNA is called nuclear membrane and adjacent to it is a highly folded membrane called the **endoplasmic reticulum**. The endoplasmic reticulum helps to coordinate the synthesis of the proteins and phospholipids. To aide in the making of the proteins, there are ribosomes embedded in what is called the **rough endoplasmic reticulum** (rough ER). There is also a **smooth endoplasmic reticulum** that makes fat/lipids/membrane. Once the protein and membrane are made they are sent off to the **Golgi body** to be put into the membrane in the correct orientation and then sent off in a vesical (small capsule) to be put into the proper place. Basically, the Golgi body is the post office, the rough and smooth endoplasmic reticulum are the factory, and the nucleus is the boss reading the book (DNA) and saying what needs to be made.

The other organelles deal with making energy and breaking down waste. Everyone remembers the mitochondria as the power house of the cell. No one remembers the vacuole that is for storage, or that the lysosome breaks stuff down and recycles it for later use. In general biology class, we spend a lot of time memorizing the structure and function of the cell and its organelles. I am a process person so once I see the process of going from the DNA to the protein embedded in the membrane, each of the organelles is just a stepping stone in the process. It makes sense and is not just rote memorization. If you take a moment to look at the cell and follow the process, it will help you to understand and remember the parts. Remembering that the endoplasmic reticulum is a folded membrane that is next to the nucleus and assists to make protein and lipids, helps to see the large picture of how cells function based on their structure. In the lab, we can grind up the cells and analyze the different proportions of the parts, which helps us to diagnose problems or identify the specialized function of a cell. But basically all the parts come together to get energy so as to make proteins do the work and then recycle the parts when not needed. The magic is how the cell knows what to make and when to make it.

Gretchen Watkins

How does the cell control its environment? Why is it important for the cell to control its environment?

4

Energy

Energy is a big part of life. Two major categories of life are: organisms that make chemical energy for themselves from the sun (autotrophs) and those that consume other organisms (heterotrophs) for their chemical energy. We think of them as plants and animals, but it is a bit more complicated when you get into the small stuff. The small microscopic stuff (prokaryotic/bacteria and protozoan) are rapidly changing, trying new mutations out, adapting to the ever changing environment. The big stuff (plants and animals) found something that works, namely the mitochondria and chloroplast, and so they protect them as specialized organelles inside their cells. My students remember the mitochondria and the chloroplast as the power house of the cell and the green stuff. The mitochondria converts the chemical energy stored in the carbon bonds of sugar and turns it into a usable form of chemical energy for the proteins to do work. This energy is called ATP. Plants are the perfect organism as they make the chemical bonds of sugar for themselves using their chloroplast and then break down their sugar using the mitochondria. We animals have to live off consuming plants and or animals that eat plants, because we have no chloroplast. This chapter will focus on how energy is gained from the sun and turned into chemical energy of sugar in the chloroplast, and how the mitochondria turns sugar into useable energy for the proteins.

Dancing with my energy levels

I am a middle aged woman and I find myself often tired, especially if I don't pay attention to my health. Low energy at the end of the work day becomes part of my normal routine. When I was young, after school/work was my time to play. I was in sports after school and when I was a young adult I would bike, run, or do anything I could find to do outside when I got home. After work/school was my most active time. Once I had kids, keeping everyone fed, washed, and happy took the place of my routine after work outside recreation. I got less sleep and my diet slipped to what I could grab on the go and so my energy waned. At this point of my life my job as a water quality engineer consisted of mostly sitting behind a computer writing reports and requesting grant proposals. It's a common story of kids, job and getting older equals to being tired or as some people say, "slowing down". But I wanted to have the energy to play with my kids, especially as they got older and more fun. I just had to figure out how to be healthy.

I was a good scientist and observed that there were people out there older than I was that still had lots of energy. So I researched healthy active older lifestyles and found there were a few things I could change. One thing was my diet. I make it sound like I was a super healthy eater before I had kids but in all honesty, I loved myself some mac and cheese and Ramen noodles. In fact, when I was in college I was diagnosed with scurvy. That's right, the thing that pirates get because they do not have any fresh food. Ramen and mac and cheese was very inexpensive and I had a hot plate in my dorm so that is what I ate every meal. It afforded me the ability to buy a ski lift ticket a few times a week. But I found I was often tired and the doctor noted I had bloody gums and so he did some blood work and determined I had scurvy. My dad decided he would get me a meal plan at the student union. So in addition to my research I had a prior personal experiment where nutrition had changed my energy levels.

To relate my story to my understanding of cells we need to look at what our bodies burn to make energy. Carbon, bound together into a simple sugar is what the mitochondria needs to make the usable energy. Noodles (examples: Ramen and macaroni) are made from the endosperm of the wheat seed, which is basically sugar. The high powered sugar is what the plant makes for the new baby plant to have a good start. I am not a baby plant with roots in the soil, and so consuming refined sugar only did not provide me with enough energy that I needed. Thus, lo and behold, I got scurvy. There must be something more to making energy. Scurvy is a form of deficiency in vitamin C (ascorbic acid). Vitamin C is a cofactor that helps an enzyme that makes collagen (part of skin tissue).

This is getting complicated so first we must talk about what an enzyme is and what it does before we can understand the cascade of events that leads to energy production.

Enzyme – the action to reaction

Enzyme is a substance that is produced by a living organism that acts as a catalyst in biochemical reactions. Catalyst is a substance that increases the rate of a chemical reaction without undergoing any permanent chemical change. To put it simply, an enzyme helps reactions happen in living things. At the most basic form, there are two types of reactions, endothermic and exothermic. The difference is that endothermic are adding energy to the compounds, and in exothermic reactions, energy is being released from the compounds. Both reactions need energy to get them going like a match to the wood. But in endothermic reactions, you will have a net gain of energy to the compounds like making or synthesizing organic compounds like sugar or cellulose from photosynthesis, whereas in exothermic reactions you have a net loss of energy from the compound, like breaking down or burring of wood to get carbon dioxide. To put simply, **it takes energy (endothermic) to build (synthesize) and release energy (exothermic) to breakdown (digest).** Similar to the fire example, living things gain their energy from an exothermic reaction of breaking down carbon compounds. Also similar to the fire example we need a match to make the reaction happen, scientist call this activation energy. In biological systems, enzymes work to reduce the activation energy and start the exothermic reaction for us to gain energy. One place we hear the word enzymes in everyday world is stomach enzymes, since they are the first step in digestion for breaking down food.

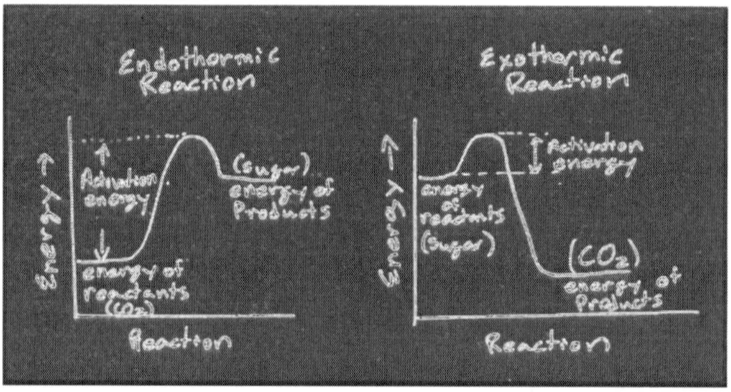

Figure 8 Graph of endo and exothermic reactions with activation energy

Enzymes are specific for each reaction. Enzymes that are in our stomach are very different than those in the mitochondria. Biological systems are very different than burning wood. In biology we need to slow the reaction down so that we do not just burn up like a fire, giving us time to capture the energy. There are many steps, each one taking energy to ultimately gain energy. These steps slow the process and allow us to make energy compounds that are useable to sustain life. The steps from the start of digestion to ultimately using the energy like muscle movement are very complicated.

We are all familiar with the first step in digestion chewing, it is also my favorite part of the process. Chewing is not just a physical process of breaking the food into small pieces. At this step there are enzymes that are delivered via the saliva. Next in the stomach, new enzymes are delivered and food is broken down so that it is small enough for cells to transport the nutrients around. One thing that must be mentioned here is the other living organisms (bacteria) that coexist in our digestive track, each with its own chemical process to breakdown food and gain energy. Some of these organisms are beneficial and work with our digestion to aid in the processes while others harm the system or throw off the balance integrity. We will come back to this coexisting with other organisms in chapter 5. For now what we need to understand is that enzymes in our stomach are specific to breaking down food and in each step of the process of gaining energy, we need not only the enzymes but also the right chemicals like coenzymes or pH to make the reaction happen.

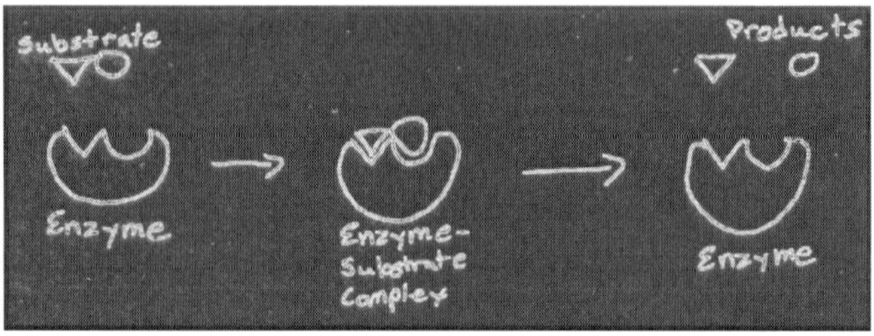

Figure 9 Lock and key diagram of enzyme mediated reactions

Digestion

Organs like the stomach make compartments for specific reactions to happen. In a cow, there are multiple stomachs having different environments for the breakdown of specific foods. Cows are amazing to me that they can get so big on just eating grass and yet we cannot. The reason they can is because they can digest cellulose. Can they really digest cellulose? It is actually their two stomach system that creates an environment for bacteria that digest cellulose that is then sent to the next stomach so that they can finish the digestion of the more simple sugar. It is their two stomach system that allows them to coexist with bacteria that can digest cellulose that give them this advantage. In eukaryotic organisms, the final compartment that allows us to gain energy from breaking carbon bonds is the mitochondria.

The mitochondria is a membrane bound organelle that floats around in the cytoplasm (gushy liquid) of the cell. Based on the fact that mitochondria have their own circular DNA (similar to that of prokaryotic cells) we believe it was once an independent organism. Evidence points to the conclusion that eukaryotic cells incorporated the mitochondria into their cells, similar to the beneficial bacteria of the cow stomach, to use the energy they create. The mitochondria are inherited in humans from the egg, in other words from the mother. Each subsequent cell received mitochondria as the cells grow and divided to become a multicellular human.

The power house of the cell was a good acquisition for the eukaryotic cells. The Mitochondria has a specialized enzyme called ATP synthase that captures energy as hydrogen move from one side of the inner membrane to the other, combining with oxygen to form water. Organisms used to respire anoxically (without free oxygen), like fermentation or glycolysis. Anoxic or Anaerobic respiration like fermentation and glycolysis has about 2 net ATP from one glucose (simple sugar) molecule. Aerobic respiration uses oxygen as the terminal electron acceptor that breaks down pyruvate into a two carbon molecule.

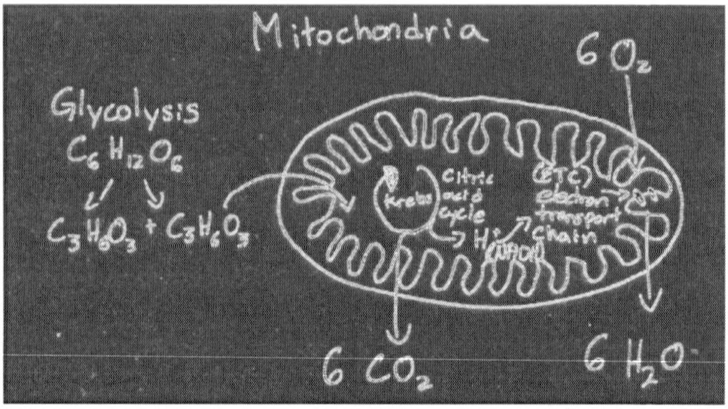

Figure 10 Mitochondria

Glycolysis – the start of sugar rush

Glycolysis is the first step to break the glucose (sugar) from a 6 carbon molecule to a 3 carbon molecule called pyruvate. Glycolysis happens outside of the mitochondria in the cytoplasm of the cell. Glycolysis does not need oxygen to happen and if there is no oxygen present, as happens in your muscles if you are running a long way, the 3 carbon pyruvate never goes in the mitochondria, instead it just gets converted into lactic acid. This anaerobic step allows the muscles to get 2 ATP, but the oxygen is needed to get the full 38 ATP with the use of the mitochondria. Therefore you must stop and take a breath for a while to diffuse oxygen back in to the muscles so you can have the energy to run fast once more. With oxygen present, respiration goes on to the second step inside the mitochondria.

After glycolysis breaks glucose into two pyruvate molecules the pyruvate is pulled into the mitochondria when oxygen (O_2) is present. The Krebs cycle is a cycle that breaks a carbon off of the pyruvate and sends it out as carbon dioxide (CO_2). The remaining hydrogen are sent to the mitochondrial membrane on energy carrying coenzyme molecules abbreviated as NADH and $FADH_2$. As a side note, I have read that some high performance athletes dope with NADH and $FADH_2$ to increase performance. This just points to the importance of energy created by the mitochondria in aerobic respiration. The hydrogen (H) that were harvested from the Pyruvate are where the majority of the energy now resides and the NADH and $FADH_2$ transports the hydrogen to the final step called the electron transport chain.

The electron transport chain abbreviated as ETC is where the specialized enzyme called ATP synthase captures energy as hydrogen move from one side of the inner membrane to the other, combining with oxygen to form water. This is why we call oxygen the terminal electron acceptor, since it accepts the hydrogen to make water or H_2O. We can think of this just like energy generation using falling water turning a turbine to create energy. In the mitochondria, the electric potential from the buildup of hydrogen on one side of the membrane to the other is analogous to the height of the water fall that ultimately turns the turbine. Then the turbine is the ATP synthase that is turned as the hydrogen move across the membrane. In the water analogy, the stored potential energy of the perched water is the same as the potential energy held by the NADH and $FADH_2$. Yet in respiration, the energy is converted into another stored chemical energy in the form of ATP. ATP is what the body can use to do work like running, jumping, thinking, and being alive.

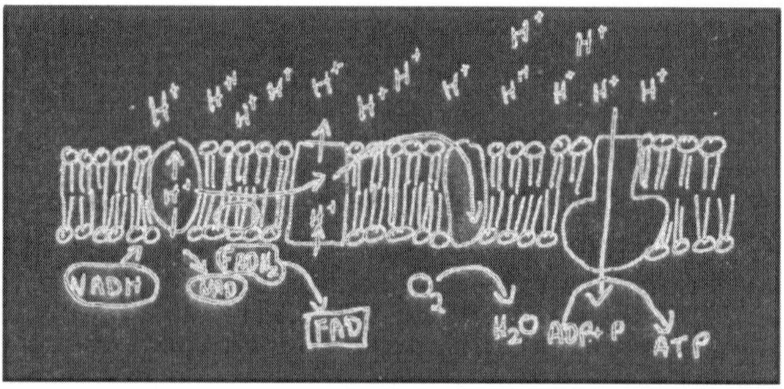

Figure 11 Electron transport chain and ATP synthase

Energy – the adjoining of chemistry with physics

Energy, an all important resource to fuel life. We all want more energy. Energy of life comes from the breakdown of carbon molecules as controlled by enzymes and movement across specialized membranes.

Since carbon is the building block for fat, sugar, and protein, organisms can gain energy from these sources depending on their specialized organs and/or organelles. Each organisms has an optimum process for this breakdown that is unique to that organism. But all life is made from carbon and uses carbon for energy. This thought process brings most of us to the question of 'where does the energy come from to build this stored chemical energy?' The answer is the sun.

Thermodynamics and Photosynthesis

Thermodynamics is the field that studies energy. There are three famous scientific laws that are established from thermodynamics. The first is the law of conservation energy that simply stated that nothing can be created or destroyed it can only be changed from one form to another. The law of conservation is critical to our understanding of metabolism. Potential energy stored in sugar is converted to energy in ATP with the mitochondria, then ATP is converted into kinetic energy with our muscle's movement. This example brings us to the second law of thermodynamics which states that entropy will increase within a system. Entropy is the degree of disorder. Life is the fight against disorder. The organization and compartmentalization of cells, their membranes, proteins, and genetic material all work together to fight against increasing entropy and disorder. In metabolism, the increasing disorder is the conversion of potential energy to heat that cannot be used by the organism. We say it is lost, but it really is the energy that radiates out of the human and into the environment and ultimately out of the world. Luckily, we have this yellow ball of nuclear power that radiates energy back to the earth, which we know as the sun.

Photosynthesis is the process that captures the sun's energy and converts it into potential energy stored in sugar. The chloroplast is the organelle that controls photosynthesis. The chloroplast takes two steps to first capture the energy and then use the energy to make the sugar. The word photosynthesis when broken into two parts tells us the two steps, photo- capturing the suns energy, and synthesis- building the sugar. The mechanics of photosynthesis are essentially the reverse reaction of respiration. Now instead of finishing with oxygen pulling the power of the hydrogen through the ATP synthase to create water; photosynthesis starts with breaking water with photons interacting in the pigment of chloroplast. Then the hydrogen are passed on to fuel the creation or synthesis of sugar in the Calvin Cycle.

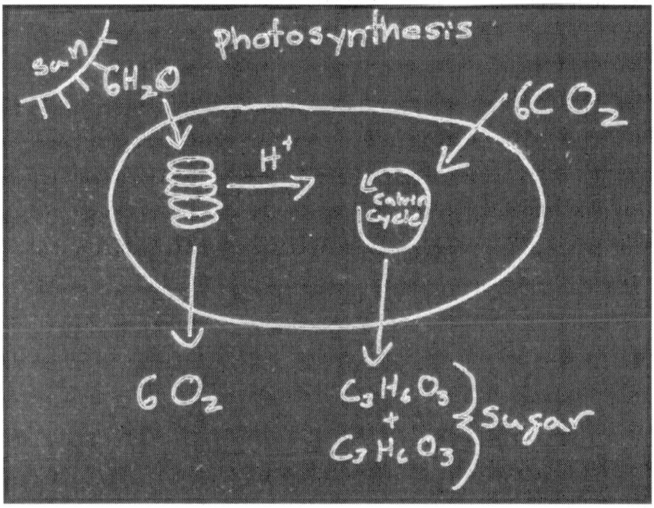

Figure 12 Photosynthesis

Chloroplasts have chlorophyll, a green pigment that adsorbs the energy of the photons emitted from the nuclear reaction of the sun. The chlorophyll is in a membrane sack called the Thylakoid. When the photon strikes the chlorophyll in the Thylakoid membrane it breaks the water apart sending the oxygen out as waste and sends the hydrogen on to power the Calvin Cycle. The Calvin Cycle takes in the carbon dioxide and connects the carbons into chains of sugar, and the cycle is complete. The sugar that is created in photosynthesis is then stored and used by the plant for energy, for growth, and for reproduction as the plants mitochondria convert the energy back into a useable form of energy called ATP. Plants have evolved over the years to store their chemical energy captured from the sun so efficiently that all of the animals' exist off of their bounty.

Energy is the connection that holds all life together. If we track and understand the movement and conversion of energy then we understand life and the connections that sustain life. So far, we have covered the molecular connections and how they come together to create different molecules (fat, protein, sugar, and nucleic acid), and how those molecules come together to create cells and their organelles. The true magic of life resides in these connections. From here we will explore how the cells are connected all the way up to the ecological connections. But remember that all life connections starts with chemical connections, from atom to atom and sun to earth.

How does the cell get energy? What forms of energy are there? Why is energy so important?

5

Communication

Communication is defined as an exchange of information. We talk, we have body language, hand gestures, and so on. Communication is complicated stuff. Humans have become as successful as we are because of communication. Generational knowledge can be passed, large group knowledge can be shared, and the individual only needs to contribute to the conversation. Collective knowledge has been expanded with the internet and media that we share. There are about 22 billion computer type devices connected to the internet and about 7.8 billion humans in the word. Our brains have about 100 billion connections. Though these connections we can explore a multitude of interests. Just like our brains, what we focus on will build connections and what we neglect will dwindle. The analogy I like to make is from weight lifting and looking at how body builders that only work on their upper body and then have these skinny little legs. Or when we are highly stressed for long period of time our higher order thinking suffers.

The analogy from the internet is how we have the ability to connect the world but we focus on one thing like politics or cats and become obsessed. Communication is about connections and the more diverse our connections the better off we are in life.

Health and stress, hand in hand

As I am writing this paragraph I find myself quite obsessed with reading about COVID 19 virus and how it might be possible to go back to school and be safe. I know logically that focusing on this one aspect of life is not healthy and I need to develop strategies to diversify my thoughts. Exercising, spending time with family, keeping a task list, and writing my book are all ways to keep my mind healthy and active. When I am being healthy, new ideas and connections are developed in my brain. The brain works by making connections between nerve cells. Development of new connections can only happen in low stress state. A high stress state is very taxing for the brain and so it cannot spend energy developing new connections. Therefor I work to keep my mind active and not stressed by developing a routine where I allow for time in my day for each activity that is good for my brain. Because what am I if not my brain and memories?

Teachers understand that routine and low stress environments are good for learning. Yet some stress can be good to push us to do what is best. For example, my routine of waking up and exercising can provide me stress when I am going to bed and just want to binge watch Netflix 'til late at night. We must use stress to push us in a positive direction. When I am reading about COVID and freaking myself out, I try and focus on learning the mechanism of the immune system so I can communicate it to students so they do not freak out and take actions that are safe. In fact the immune system is a wonderful example of cellular communication.

Cellular Communication

The immune system works in stages. First our bodies try to keep germs out of our bodies with skin, mucus, hair, and other protection mechanism. But this is not foolproof so next our bodies need to recognize an invader, if it enters our blood stream, different than our own cells. This recognition is cellular communication. The immune cell recognizes other cells by proteins imbedded in the cellular membrane. Remember the membrane is outside of the cell. The protein that is stuck to it can be recognized by another cell. Our blood cells have protein that give us our blood type. Blood type A has an A protein and B type has a B protein. If you have blood type A and you get a blood transfusion of B blood type your immune system will recognize the B protein as not yours and attack the blood that was meant to help you. This same system will recognize foreign bacteria and viruses. The image below depicts the recognition of a foreign blood type by an immune cell.

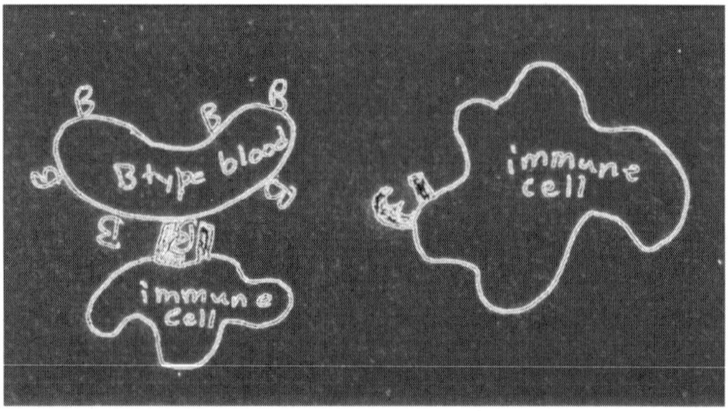

Figure 13 Blood

Communication does not stop with recognition. The purpose of communication is to trigger a response. In the immune system example the response could be for the immune cell to destroy the foreign cell. In the human body the other two major communication cells or systems of cells are the nervous and hormones. Communication follows the same steps of recognition, to transmitting the signal, to a response. Nervous cells are quick responses like pulling your hand away from hot pan, or closing eyes to bright light. Hormones on the other hand are slower responses but they last longer, for example signaling the start of puberty. Each of these types of cell communication follow the steps of receiving the signal, transmitting the signal, and finally responding to the signal.

Citizen Science

When I was working as a water quality specialist for a tribal nation in northern Wisconsin there was a large mineral deposit that was being explored near the reservation. The tribe fought the companies looking to exploit the minerals. The tribe was small and had few resources to protect their reservation so they had to figure out it even though few of them had a scientific background. The amazing thing was by the end of the decade long fight, the tribal members became amazing scientists. Talking to people about the COVID 19 reminds me of how people with little science background can learn when faced with a crisis.

The spike protein on the COVID 19 virus is a term many people are familiar with now. People know the specific spike protein for COVID 19 has to do with how it connects to our cells to infect the cell it connects to. People also know the spike protein has to do with the vaccine. This understanding leads to the understanding of cellular communication and immune response. The spike protein is specific to COVID 19. Our immune system does not recognize it as bad yet as it is a new mutation of a class of virus. Vaccines work by training our immune system to recognize the spike protein as bad so it destroys the virus. The immune cells that recognize and remember the COVID 19 spike protein actually remember it based on the DNA making a protein that is the counter part to the spike protein. We will get into how DNA codes for proteins in the next chapter but for now just know that it does. Therefore the vaccine is essentially programming your DNA to find and kill COVID 19.

Cellular Differentiation

Communication is also very important in cellular differentiation. Cellular differentiation is what happens when one cell is different from another cell even though they both have the same DNA. It is easiest if we start at the very beginning of life. When we are conceived, a mothers DNA merges with the fathers to create a unique set of DNA we call that new cell a zygote. All subsequent body cells will get a full set of the DNA. The DNA has all the information that makes you, you. And each cell has all of the information. Yet only part of the DNA is read depending on what type of cell it is, skin or muscle or nerve or whatever cell type it is. The analogy I like to make is that DNA is the book of life and differentiated cells are programmed to read only a specific chapter. Skin cells read skin chapters and muscle cells read muscle chapters. Cellular communication is how they know what chapter to read.

Once we have a new zygote the cell begins to divide following the cell cycle of growth, DNA replication, more cell growth, and then dividing through mitosis. During the growth stage the cell is making enough organelles and cell parts so each new cell gets enough. During

DNA synthesis the cell is replicating a copy of the DNA so each new cell will get a full copy of the DNA. Then the cell grows more in preparation for **mitosis**. Mitosis is the division of the DNA and nucleus that holds the DNA. Mitosis follows the steps of prophase where the chromosomes condense, metaphase where the chromosomes line up on the middle of the cell, anaphase where the replicated DNA are pulled apart, and lastly telophase where the nucleus reforms and the cytoplasm splits creating two new identical cells. These new cells are undifferentiated for about the first week of embryonic growth. The embryo is basically a ball of cells that all look the same and can become any type of cell. These undifferentiated cells are called **stem cells**. The ball of cells called a blastocyst starts to imbed in the uterine lining and the uterine cells communicate with the embryo cells to start cell differentiation.

As the ball of cells grows it starts to get too big to diffuse the gases like oxygen and carbon dioxide so the first cells that differentiate are ones that create blood flow for the exchange of gases. From this point on the cells differentiation is based on position. The cells are connected to create more complex tissues like skin, muscle, and nervous. These tissues come together to create organs like heart and lungs. Then these organs come together to create organ systems like the digestive system and cardiovascular system. This order of organization is how complex systems can be so large and still operate together. Communication is the key for these systems to work and it starts at the cellular level. Each cell communicates its location to the adjacent cell. So the top cell will be skin so on and so forth.

The body layout is the most sensitive time of development for the human. It happens in the first and second trimester. The cells can signal each other in a variety of ways. The cells can have direct connection or send messages from a distance. Hormones are chemicals the body uses for long distance communication. Remember protein embedded in the membrane was one way to receive a message and the way that works is a lock and key type chemical attraction to a specific signal. Some hormones can diffuse directly through the membrane and chemically interact inside the cell to trigger the message. Some chemicals we consume or breathe can be similar to certain hormones and disrupt the process. During this early time of development, getting the correct signal is critically important.

Communication is first about receiving the correct signal but next is how it is transmitted to create the response. In embryonic development the response is reading the specific part of the DNA that will create the type of cell needed. Many times the cellular response is to read a part of the DNA so specific proteins are created to do a certain job. For example, when you drink a lot of milk your body is signaled to mate the protein enzyme lactase so you can bake down the lactose in milk. These processes can be turned on and off with feedback loops depending on what needs to happen. Normally if something needs to be made or synthesized then the signal that will turn it off is the product. If something needs to be broken down then the chemical that needs to be broken down is the signal to turn the pathway on. Understanding these pathways is very complex and the basis for much research.

The complexity of communication is amazing. To think that our thoughts and actions are based on chemicals interacting in our body makes me think how complex we are. When I am communicating with the virtual assistant Alexa, I am both amazed that she understands me and amazed when she does not. Organization and building communication pathways that are positive creates a functional organism. Yes it is simple as electrons moving from one to another to turn on or off an action.

What is heredity? How does the cell transfer information? What is diversity and what processes help to create more diversity? What do mutations lead to?

6

Gene Expression and Regulation

Gene Expression is the process of reading the DNA to make proteins. In chapter 5 we talked about how the cell is triggered to read a section of DNA. This is the first step in the regulation of gene expression. Only one percent of DNA is a direct code for a protein. The rest is regulatory or non-coding DNA that is no longer used. This chapter we will talk about DNA, how it is expressed into proteins, and how the regulatory parts control the expression.

Let's make proteins

We will start with what is DNA and how it is put together and copied. One of our macromolecules that we did not talk about much in the beginning of the book was called nucleic acids. Nucleic acids denote the chemical makeup of the building blocks of DNA and RNA. DNA is made of 4 different nucleotides, adenosine, guanine, thymine, and cytosine. These 4 compounds make up the code of life. When I am presenting to students about why they should sign up to take AP biology, I tell them they will learn the secret to how just 4 chemicals make every living thing in the world. Think about it, 4 different chemicals make up all the diversity from a worm to a human. All living things have the same 4 letters.

So here is the secret. It is complicated so I use the analogy of cooking to help make it understandable. I have a cookbook that was handed down from my grandmother. When my mother was teaching me to cook, we were not allowed to have grandma's cookbook out on the counter while we were cooking. This cookbook had every possible recipe in it. The cookbook was kept safe. Just like DNA is protected in the nucleus. We had to make a copy of the recipes on a little note card that we would use when cooking. In the nucleus DNA is copied into messenger RNA. RNA is basically the same thing as DNA but instead of thymine it has uracil. And instead of being double stranded it is single stranded. The process of copying is called transcription. Next we take this note card and use it to make the food.

Figure 14 Transcription

The kitchen is a messy place and there are lots of ingredients and cooks. First there is my mom, the master chef analogous to the ribosome (rRNA) that reads the code. Then there is me, the helper that brings the ingredients, analogous to transfer RNA. The ingredients are amino acids, the building blocks of protein. So it goes, my mom reads the note card, shouting out first ingredient methionine (always the first amino acid). So I transfer the methionine to her and we begin what is called translation. Next she shouts leucine, and I transfer leucine and she attaches it to methionine. So on and so forth until we have a long chain of amino acids that ten gets folded up into a protein.

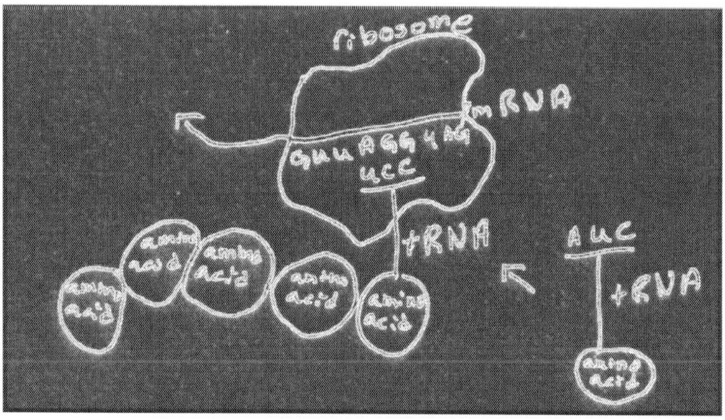

Figure 15 Translation

Transcription and translation – the recipe

So think of the cooking analogy as you read the steps of gene expression. First step is **transcription** where the DNA is copied into messenger RNA (mRNA) and sent out of the nucleus to the ribosome. Then in **translation** the ribosomes read the mRNA code and the transfer RNA (tRNA) brings the corresponding amino acid. Each code is 3 nucleotides long, we abbreviate each nucleotide as their first letter. DNA has the letters A, T, C, and, G and RNA has A, U, C, and G. The code is read based on hydrogen bonding. Hydrogen bonds are similar to covalent chemical bonds in that the elements are sharing the electron, but hydrogen bonds are not as strong. The two strands of DNA are stuck together with these hydrogen bonds between the nucleotides. It is based on the number of hydrogen bonds each pair makes and they are called complementary pairs. For DNA, A pairs with T through two hydrogen bonds and C pairs with G through three hydrogen bonds. For RNA A pairs with U and C pairs with G. In translation between DNA and RNA, the T of DNA will complement to an A, and an A of DNA will complement to a U of RNA. Students love decoding the mRNA sequence to find the amino acid sequence that will make up the specific protein. The triplet code of mRNA binds to the code of the tRNA to bring in the corresponding amino acid. There are 20 amino acids used in living things and proteins can vary between 50 to 2000 amino acids that come together to make the chain. This is the process of the protein expression and is seen modeled in the following image.

Just like in cooking if you substitute an ingredient your meal will taste different, in gene expression if you have a different amino acid you will have a different protein. The process starts with the DNA code, and the code is copied into all the cells in your body based on the complementary pairing of A to T and C to G. This pairing is based on hydrogen bonding, therefore chemicals, radiation, and other substances can cause the copy to be miss read. This results in a mutation. Sometimes the mutation can be permanent if it happens to your egg or sperm. Most times the mutant cell is just absorbed by your immune system. Sometimes the mutation can result in cancer and if the cancer is not recognized by your immune system then it can multiply and become a problem.

An example of a mutation that has been passed down for many generations is sickle cell anemia. Sickle cell anemia is from a single mutation in the DNA where one letter was off and therefore one amino acid became off. This results in the protein that makes up the red blood cell to be mutated, and so the red blood cell is now sickle shaped and has a hard time carrying oxygen. The interesting thing is, sickle cell anemia gives an individual resistance to malaria. If the only gene expressed is the one forming sickle cell then it will be fatal. With sickle cell anemia, the gene can be expressed from either parental genes you inherited. You get one from your mom and one from your dad. If one is dominant and the other recessive then only the dominant one is expressed. The recessive one will be wound around proteins so tight that it can't be read. In the case of sickle cell both normal and sickle cell genes are expressed so neither is recessive. Therefore if the individual has one normal and one sickle gene then their blood will be half normal and half sickle.

This gets into how we control what is being read or not read. We learned we can receive a signal that tells the cell what to do. Since we have two genes that code for the same trait, the first thing that happens is the recessive traits being wound tightly around proteins so they cannot be read. This also happens in differentiated cells where the parts of the DNA that are not needed for that cell type are wound tightly around proteins. Next, there are transcription factors that are proteins that assist in reading the DNA into RNA. These transcription factors can be coded in other parts of the DNA but they come together to read the specific segment of DNA that codes for the protein. These transcription factors can be blocked or enhanced depending on the signal received by the cell.

Bacteria or prokaryotic organisms replicate with the mitotic process by creating a clone of themselves. The DNA is the same in each cell for the most part. Bacteria have a small circular piece of DNA called a plasmid that is used to transmit genes that help other bacteria. Viruses can also manipulate the DNA of an organism. Some virus like herpes will recombine their DNA into the host organism so the host cell will express the viral code to create new viruses to spread to new hosts. Once scientists figured out that there was a way to move DNA into a cell and make copies of it so that the DNA would be expressed, they immediately started to use these tools for their medical advantage. The most amazing example of this recombination was when scientists moved the gene for insulin and spliced it into bacteria so that the bacteria would create human insulin in a test tube. This was much better than extracting it from pigs as they did in the past since many people had allergic reaction to the non-human DNA.

Genetic engineering or biotechnology is the field to use these mechanisms of bacteria and viruses to recombine DNA with DNA that could correct an illness or improve an organism. CRISPR is a new tool that makes this process of recombination much more useful. CRISPR stands for "cluster of regularly inter spaced short palindromic repeats" as these are the area of the DNA that can be recognized for splicing DNA. Genetically modified products are much more mainstream since CRISPR's discovery.

How do cells have all the same DNA in a single human but some cells look like skin and some look like heart? How do we genetically modify organisms?

7

Heredity

Heredity often makes me think of the purpose of life. I hear many people talk about children and their legacy and how descendants are their purpose. Definitely having children has been enjoyable and fulfilling for me as an individual, and survival of the species is contingent on offspring. So I think propagation is important for the continuation of life indeed. But then I start thinking about it in a larger context of all living beings. What does it mean to be a specific organism? The dictionary definition of heredity is: the transfer of genetic information to offspring. This simple transfer of DNA to a new individual only happens in asexual reproduction. Like when a spider plant breaks off a piece of vegetation and a new spider plant grows. Or how bacteria replicate based on replication. But even with the simple bacteria cloning there are ways to introduce diversity.

Genetic diversity

Diversity is actually what helps life persist through an ever changing environment. Therefore species that replicate based on creating unique individuals helps the species survive. Sexual reproduction is the process of transferring half of the DNA from one individual and combining it with half from another individual to create a unique individual. This process ensures there will be a diverse population so if something like malaria comes along, some of the individuals in the population might be resistant to malaria as seen in the example in the last chapter.

Fertilization – the fusion of gametes

There are three major ways sexual reproduction leads to genetic diversity. All of them happen during the process of creating sex cells known as gametes or sperm and egg in animals, and pollen and ovum in plants. First the cells have to become half of the amount of DNA so that when they join with another gamete they end up with a whole set of DNA. Remember each cell is made up of half the DNA from each parent. In humans, we have 46 chromosomes; 23 of them we get from our mother and 23 of them we get from our father. **Meiosis** is the process for making gametes; it follows the same process as mitosis but it goes through nuclear division twice without replicating the DNA in between. The first splitting of the DNA is called meiosis 1 and the second is called Meiosis 2.

In meiosis 1, the homologous chromosomes separate, and in meiosis 2 the sister chromatids separate. Homologous chromosomes are the chromosomes that code for the same trait like eye color. From the pair of homologous chromosomes, you get one from your dad and one from your mom. So if your mom has blue eyes and your dad has brown each homologous chromosome codes for eye color but you might have one that codes for blue and one for brown. During meiosis 1, the homologous chromosomes pair up and portions of the chromosomes can crossover to result in a chromosome that is a mix of genes from your mom and dad. Crossing over is the first step in building diversity. Next the homologous chromosomes line up to be pulled apart. The chromosomes independently assort so some from your mom and some from your dad will separate during meiosis 1. You have 23 pairs of chromosomes and there is no telling if a gamete will get a certain proportion from your mom or dad. Crossing over and independent assortment results in creation of sex cells that are completely unique. The final step of sexual reproduction and creation of unique individuals is joining the gametes called fertilization. Fertilization is random, this adds to the diversity of the individual called a zygote.

Of Punnett Squares and Heredity

Since the new individual receives only half of the chromosomes from each parent, then there is a 50 percent chance of getting a specific trait from a parent. There are two parents you are receiving genes from, and if the traits follow Mendel's genetics then only one trait will display. Mendel is the father of our understanding of genetics. He studied true breeding traits of pea plants. He controlled the breeding and kept meticulous notes so he could apply statistics to his investigation. His studies proved that traits randomly separated and recombined during sexual reproduction, even before we could see the chromosomes separate microscopically. The statistics Mendel describes in his work were expanded on by Punnett work of the Punnett square.

The Punnett square allows us to organize the traits so we can predict the outcome of a crossing. If the individual donating gametes is homozygous then both traits the individual has are the same. We denote the dominant trait as a capital letter and recessive trait as lowercase. If the individual's genotype is heterozygous, then they will have one dominant and one recessive allele showing the dominant trait. The terminology is important to keep track of the genes. Because we get two homologous genes that code for the same thing, we need to use statistics to find the probability of a specific outcome. The outcome or what it looks like is called the phenotype. The phenotype is really what gene is being expressed, as we learned about last chapter. The gene is the part of the DNA that is being expressed. Since you get two genes that code for a specific trait we call one gene an allele. So if both alleles are the same then you are homozygous and if they are different you are heterozygous for that trait. Example of a genetic probability question would be: if you have a blue eyed person crossed with a brown eyed person what is the probability they will have a child with a blue eyed phenotype? First you can say that the parent with blue eyes is homozygous recessive because if they had either allele dominant then that would have displayed. But the brown eyed parent is more difficult as they could be heterozygous or homozygous dominant. The way to know is to figure out if one of their parents has blue eyes or if they had a child that resulted in blue eyes. Looking at the two Punnett squares, you could see that they could have a zero possibility of a blue eyed child if they are homozygous and a 25% chance if they are heterozygous.

This is where I usually get the question, "what about green eyes?" Here is where the fun begins. Mendel only understood inheritance of traits that were fully dominant or recessive and where there was only one gene that codes for the phenotype. With our advancements in molecular biology, we now know there are many traits that have incomplete or codominance. Also there are traits that have multiple genes that come together to express the phenotype. Eye color is one of those more complex genes. The mechanism is still the same, it just becomes more complex. In codominance, the allele being expressed is random, as seen in a roan horse. In incomplete dominance, both alleles are expressed giving the organism a blended phenotype. This is how you get green or hazel eye color.

Gene expression is how your DNA is read into proteins that make the unique individual that you are. Diversity is based on unique DNA being passed from generation to generation through sexual reproduction. Yet we are all made from the same 4 building blocks of DNA. The code of life is simple yet complex. If we break it down, we can figure out how something works and communicates and then we can utilize this knowledge to fix problems of the world. In a year we were able to figure out the protein that bonds with our cells that caused COVID-19 and from that we figured out the RNA code and created a vaccine that was made of RNA code that when injected would trigger the immune system to figure out the bad protein and build antibodies.

Heredity in Viruses and Bacteria

Viruses and bacteria are a bit different than plants and animals in how they replicate and share DNA or RNA. The existence of bacteria have been known about since the 1670s when they were first observed with a microscope. Bacteria are prokaryotic, meaning they have no nucleus or membrane bound organelles. They are much smaller than eukaryotic cells and so were not as well understood at that time. In reproduction, they simply copy their DNA and split into two. It was not until the 1920's that we learned how bacteria share DNA to increase diversity. Bacterial transformation is the movement of genetic material from one bacteria to another. Frederick Griffith observed that bacteria which were killed were then able to transmit traits. They did not fully understand the mechanism of transmission. After years of studies by multiple scientists, we now know there are small circular pieces of DNA called plasmids that bacteria can share. We know that there are enzymes the bacteria use to splice up DNA and copy DNA into their code. In AP biology we use transformation to make a bacteria that does not glow, so that they glow.

Viruses are so small, we were not able to see them until X-ray crystallography microscopes were developed. Rosalind Franklin took the famous picture of the DNA double helix that Watson and Crick used to publish in *Nature* explaining the structure of DNA based on the chemical bond length of the nucleic acids. Sixty years later, scientists investigated how bacteria fights off viral attacks and have discovered how to use the bacteria's adaptive immune response to revolutionize gene editing. It is all about the DNA and protecting the DNA that make the organism who it is. The mechanism of the virus is to insert their nucleic acid (I use the general term as it could be DNA or RNA) into an organism so that the viral code is read to make more viruses. A virus is just nucleic acid inside a protein coat. They do not have ribosomes to read their code to make their protein. So, they insert their code into a cell and the cell's ribosomes will read and make the protein for the new virus. Now that we understand this process we can manipulate it to insert nucleic acids that code for what we need or want, like insulin.

As amazing as all of this is, that we can manipulate the code of life based on our understanding of cellular processes, I am also reminded of the complexity of it all and that we have so much more to learn. My husband was surprised that it took so long for us to develop a vaccine for COVID. I, on the other hand, was surprised how we did it so fast. Biological advancements make us think that we can play God sometimes and save the world. But the one thing I have learned about science is that everything that we know today will be different tomorrow as we learn new things. Life is complex.

What makes you who you are? How can every cell in your body have the same DNA but look different like skin or muscle?

8

Variation and Natural Selection

To understand variation and natural selection, we must start at the very beginning. Or what we think is the very beginning of life about 3.5 billion years ago (BYA). It was a really long time ago and the evidence for different chemicals on earth and how they became organic chemicals that can control their internal environment is limited. We have lots of fossil evidence that starts about one billion years ago. But before that, we just looked at the rocks and how the chemistry of the earth changed. For instance, we hypothesize that photosynthetic organisms arose because we see a lot more oxygen in the rocks of that time. We can also look at how things work now, and the DNA history in organisms that exist today. For instance, the mitochondria and chloroplast have been around for a long time with little changes. They have their own circular DNA and ribosomes as do free living prokaryotes. **Natural selection** is all about things that work out, eventually sticking around. Mitochondria, the powerhouse of the cell was once a free living organism that was consumed by another organism but they pumped out so much ATP that the consuming organism did not digest the mitochondria but instead protected it inside its membrane and allowed it to grow and replicate as the organism grew and replicated. This theory is also true for chloroplast, the organelle that can capture energy from the sun and turn it into sugar.

The wonders of the scientific method

How amazing to think that there was once a free living organism that was so cool another organism kept it alive inside itself so it could live off the extra energy. How can this be true? How can we think this? Science is based on observation and evidence. The scientific method starts with an observation where someone said to themselves, "Wow that is cool, I wonder how that works". Think about the first person to uncover a T-Rex skeleton. Then think of all the ancient people around the world. Seems coincidental that in Asia they have dragon mythology that looks a lot like the dinosaurs of their area. This is the thing, it is not just one finding or one thought. It is many findings, and communication about the findings, that lead us to make conclusions, where we then find new things to be amazed with. The advancements in our ability to observe the world only helps this process along. Now we can excavate the fossils that are microscopic, and record the date when the fossil was formed. We put this all together by communicating what we found and voila, we have our theory of evolution.

Theory and scientific law come from the scientific method and how scientists decide if something is true. Since science was around, there has been apprehension about what scientists have been concluding. It is actually in the scientific method where we want people to be skeptical and see if they can prove us wrong. Because if they can't then that helps to prove that the scientist is correct.

I bet there are people out there today that still believes in dragons. Avatar the Last Air Bender TV show had me believing in dragons. But I digress.

The levels of understanding in the scientific world start with hypothesis, this is a guess of what is happening based on prior knowledge of our observed phenomenon. In the scientific method, after we make our hypothesis we test it and collect data. The data is what we call evidence. The data is organized into quantitative (numbers), or qualitative (descriptions) measurements. Then the scientist analyzes the data and explains what they think happened.

Now this is not a theory yet. In general terms, theory means a guess. In the scientific world, you only get to theory once other people have run your experiment and found similar results. Lastly you get a scientific law. This is when you have an explanation for your theory and when people test it in different ways the explanation holds true. Like the laws of thermodynamics, gravity, energy and so on. Many times these laws have corresponding equations like $E=mc^2$ for energy's relationship to mass and speed of light.

The law of evolution builds off our understanding of genetics and inheritance. Much of the work of Mendel has been built on as a null hypothesis that states no change over time other than the mixing of the dominant and recessive traits. Null hypothesis is like the opposite of what you are testing. No change from normal conditions in Mendelian genetics would result in just a reshuffling of the genes through sexual reproduction. For example, the Punnett Square, where you have a brown mouse (dominant) and white mouse (recessive). Say the starting population has 50% of the alleles dominant and 50% recessive then you would just mix them up between homozygous dominant, heterozygous, and homozygous recessive, where the recessive are the only ones that show up as white. In the population, you might have generations with many more brown than white, but always just reshuffling the 50/50 genes. You can see this with our understanding of Punnett squares.

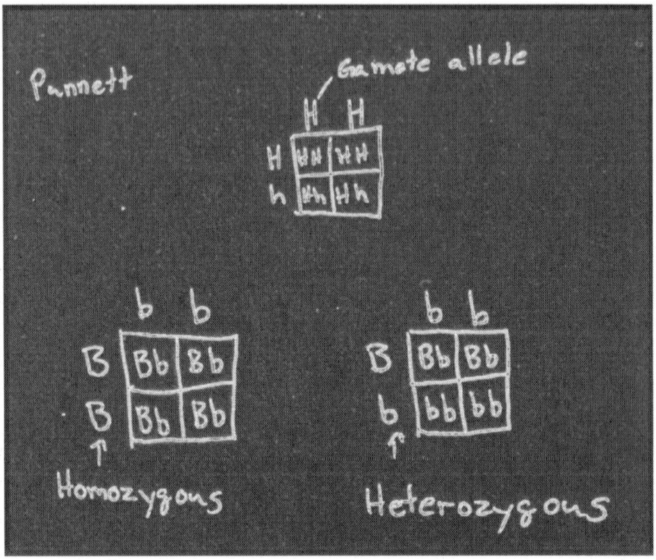

Figure 16 Punnett Squares

When we test this over and over, Mendel's work holds true under certain conditions. Mendel was a Christian monk that lived in a monastery where he gardened and studied math. He could predict what color a pea plant would be based on the genetics of the parents. I always envision him calling his other monk buddies to see him do his work and saying "what do you think will happen if I cross pollinate these two purple (dominant) flowers?" Then he would say I bet I get 25% of them to turn out white (recessive). Then all the other monks would say, "no way", but Mendel knows that the parents are heterozygous. Therefore, he is correct and gets 25% of the flowers white. Then Mendel is all "see boy's I told you so, genetics!" (mic drop).

But this would not happen unless he was controlling population size, pollination, and growing conditions so that the white and purple would have the same chance at survival. What if Mendel plucked the white flowers each time and killed them generation after generation?

Mutations and Artificial Selection

Artificial selection is the process of humans choosing a trait they prefer. We have been doing it for years with agriculture and animals. Dogs are my favorite artificially selected animals and apples are my favorite plant. With artificial selection you do not get a reshuffling of the genes, the percent of genes change over time. Evolution is the change over time. Artificial selection is a process where humans speed up the change over time. If we look at this in a much longer time frame then we also have mutations that arise in a population that could be selected for or against.

Mutations happen in the DNA and some end up being good changes. These good changes lead to reproductive success. "Good" change is the hard part to understand in this process of change. What is "good" and what is "bad"?

The story I tell my students is of a man from a small village long ago who had a son. One day the son goes off and finds a herd of wild horses and brings them back to the father. Everyone from the village told the man he was so lucky to have such a good son that provides him horses. The man responded, "Good luck, bad luck, who knows". The next day the boy was training the horse and one of them bucked him off and he broke his leg. Everyone from the village told the man he was so unlucky now his son will not be able to help on the farm. The man responded, "Good luck, bad luck, who knows". There was a war going on in the country and the next day the military came to collect all able bodied men of the town to go to war. Everyone from the village told the man he had such good luck that his son did not have to go to war. The man responded, "Good luck, bad luck, who knows".

This story illustrates that we have lots of traits and only the situation that you are in can dictate if it is going to be selected for or not. Take the brown and white mice. If there is a hawk that eats the mice and the mice live on a brown landscape, then it is good to be brown. If it snows then it is good to be white. Therefore variation in a population helps it respond to change so a population is not taken to extinction when the ice age comes. Variation comes from mutations and the mutations are passed down to offspring. This is the process of evolution.

Natural selection was the process that Charles Darwin described in his book 'On the Origin of Species' published in 1959. He described a process of descent with modification. At his time, we did not understand DNA and how mutations arise in a population. Now we study the rate mutations occur in different organisms under multiple conditions to the point that we can predict or induce the incidence of mutations. From our understanding of molecular biology and how the nucleotides come together to make DNA and how DNA is transcribed and translated into the working protein of the cells, we can then track mutations in populations over time and back into history. Extinction (loss of species) and speciation (new species) rates are displayed on evolutionary trees.

Before our ability to read DNA, we determined speciation based on what the organism looked like and if it reproduced with the one it did not look like. A polar bear is big and white and a brown (grizzly) is brown. The standard definition of species is that they are reproductively separate. Many things can keep them separate including the fact that they both just don't live in the same area. Evolution in the present day can now be seen as the extinction of the polar bear occurs while our earth warms. As the habitat for the polar bear decreases, the habitat for the grizzly bear is moving farther north and the two bears are no longer separated. Currently, as these bears come together they are mating and creating viable offspring. Our term for this is hybrid, meaning a little of both species. But as time goes on and we stop seeing polar bears, are they extinct if they mated with grizzly bears and their DNA is carried but not expressed? One way it could go is that the hybrid becomes the most successful and the brown and grizzly bear becomes extinct. Or you just have a new hybrid zone and the three bears become distinct, or something in between.

Evolution is gradual change over time. Our fossil history makes it look like we have very distinct or punctuated changes. Sometimes when the climate changes very rapidly or there is a catastrophic event killing off many species then there will be an abrupt change in the distribution of species. One thing that is important is for us to recognize the process of change so we can plan the future.

What drives change in species over time? What limits change in species? What is the evidence that shows the process of evolution?

9

Community Relationships

With the foundation for how organisms work, we can explore community interactions to find the balance in life. We learned that all life starts with chemistry and the 4 major biological macromolecules, and that they are made up mostly of Carbon, Hydrogen, and Oxygen. Fat and sugar are made of these three elements and then nucleic acid (DNA) adds up in phosphorus, and protein adds in phosphorus and sulfur. Knowing this and the properties of the macromolecules is the first step in tracking what is happening in an ecosystem.

My master's thesis was working on the drinking water system of New York City. In 1997 New York City had a system of aqueducts that transported water from lakes and reservoirs outside of the city. The farther you moved away from the city, the cleaner the water was. Some of the lakes had no homes along their shoreline and the others closer to the city had many homes. It was brought to light that the water from the lakes that were receiving the most pollution was going to the poorer neighborhood and the water from the undeveloped lakes was going to the wealthy neighborhoods. The Environmental Protection Agency (EPA) then required that NYC start mixing the water before it sent it to the different neighborhoods. This resulted in a project to figure out how to get all the water clean. My graduate project was to help figure out how to clean the water without filtering it as it was cheaper to keep the water clean instead of filtering it at the end. Specifically my project was to track phosphorus in this system.

Phosphorus is needed for life and it is one of the elements that can limit growth of life as it is not found in the environment in as high of quantity as you find carbon, hydrogen, and oxygen. Just like if you remove light from plants they die, if you remove phosphorus, they will not be able to survive. What was making the drinking water dirty was the growth of algae in the lakes. The lakes had abundant light and water so the next thing to limit their growth was phosphorus.

Development like homes and roads contribute phosphorus to the watershed. Sometimes you have so much phosphorus in a lake system that you grow algae to the point that you have green film across the surface killing off the photosynthetic organisms below. These algae and plants die back and sink into the muck where the consumers (animals and bacteria) convert the sugars from the plants into energy consuming the dissolved oxygen in the water. When this happens the lake system is very far out of balance and cannot sustain fish and larger animals that need to breathe the oxygen in the water.

The NYC water systems were not aware at this point but they did have too much algae to the point that it was making the water carcinogenic when chlorine was added to disinfect the water on its way to homes. So how do you solve this problem? From our background research, we knew that phosphorus was the limiting nutrient to the algae in lakes. We also know that development was a source of phosphorus. Our general hypothesis was that if we add less phosphorus to the lakes then less algae will grow making the water less toxic, since phosphorus is the limiting nutrient to the algae in the lakes.

Once the experimental procedure was identified and the independent variable (phosphorus) and dependent variable (algae) were set. Each of the graduate students could study different aspects to the large experiment. The results of the project found that phosphorus can be controlled if you reduce development, convert individual home septic systems to centralized systems outside of the watershed, and install storm water protections like storm gutter protections that collect sediment from roads.

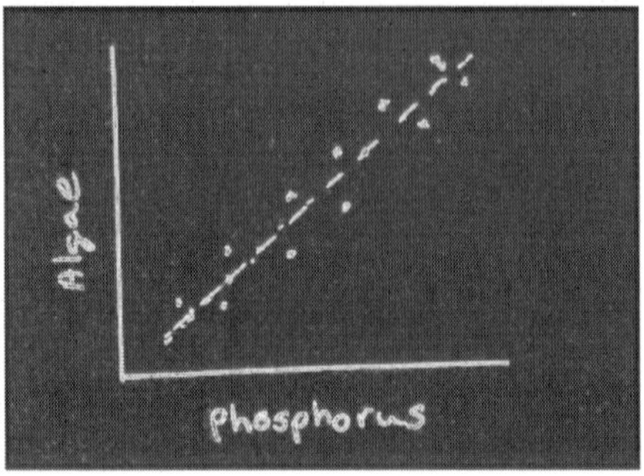

Figure 17 Generalized relationships between experimental and actual phosphorus loading versus algae growth

Finding this balance made the NYC drinking water safer. It was also less expensive to deal with the problem at the beginning of the process then the end. Keeping water clean is cheaper than filtering it at the end. Just like our human excretory system needs to keep poisons like drugs and alcohol out of our system, which is better than having our kidneys cleaning out blood for us all the time.

Identification of your independent and dependent variables is key so you can find the relationships where there is balance. The independent variable is the one you have control over and can manipulate and the dependent variable is the one that results from the change in the independent variable. Then one just has to examine the experimental results to find the balance. The wealthy people living on the drinking water reservoirs can afford the changes to protect the water for all.

Each of these investigations tend to revolve around the population of a certain organism and the interplay between an autotroph (makes their own food like photosynthetic organism) and consumer (eats others for food like the grizzly bear). When there are unlimited resources the organism is going to go through exponential growth, as the resources are limited the growth will slow down until it is even with birth and death rate. This balance is called the carrying capacity of a system. An organism can then move into a new system and start exponential growth again. All organisms follow this type of population dynamics.

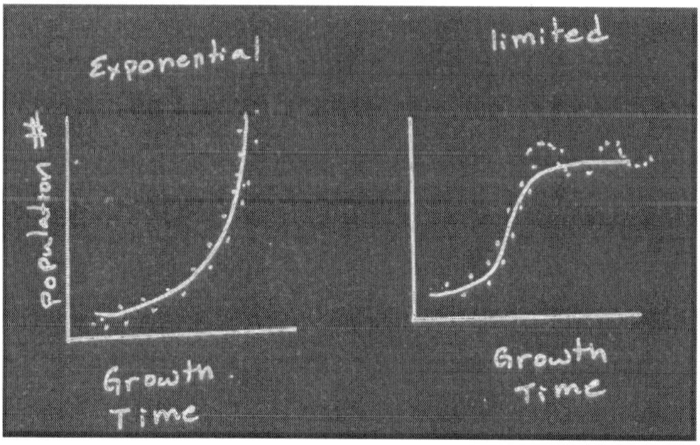

Figure 18 Exponential and limited population growth

Predator prey interactions result as the resources for the prey oscillate with prey being eaten by the predator. This is the difference between top up (limited nutrients) and bottom down (killed by predator or parasite) population dynamics. Both need to be examined to understand effects of population density rates. Naturally a boom and bust cycle occurs changing the rates of population growth. Population rates are calculated based on the number of births minus the number of deaths over time. Human population growth rates have been increasing exponentially since the advances in agriculture, industrialization, and medicine have allowed humans to increase the birth rate and decrease the death rate. Wealthy nations have figured out that population control is important so we can sustain our existing population, and thus many wealthy nations are staying steady in their population growth. Similar to the population growth rate of the algae in the drinking water reservoirs we can figure out how to find balance in other species' populations.

Community ecology is then understanding how each of these communities interact with their biotic (living like other organisms) and abiotic (non-living like nutrients, sun, water) environments. Community ecology is very complex as now we are not only assessing just one organism or an ecological relationship like predator-prey. Now we look at a full community. To do this, scientists need to organize the organisms so we can categorize them to understand the role they play in the ecosystem.

Starting with one to one symbiotic relationships, we have 3 types. First is the one that benefits both organisms like the bird that eats the bugs off the cow, which is called mutualism. Second there is commensalism, where one benefits and the other is neutral, like a dandelion seed catching a ride on some dog fur. Lastly we have parasitic where one is harmed (usually defined as the host) and the other benefits. Everyone knows parasites and relates to these relationships with having a tic on them, but often other two symbiotic relationships are overlooked. People also confuse parasitic symbiosis with predator-prey relationships. Symbiosis is specifically defined as a long term relationship so the predator eating the prey is just an ecological relationship.

When we go up another level of ecological relationships we think of food web and food pyramids. The word trophic refers to eating or nutrition so if we are referring to the trophic pyramids, it means we follow energy flow through a system. For example, let's take North American Temperate Forest, and create a trophic pyramid and food web.

It is nice to define an ecological boundary so we can account for the major players in the ecosystem. So starting with the trophic pyramid, we start with autotrophs on the bottom as they capture the energy to feed into the system. Our forest would have trees, grasses, and shrubs as our autotrophs. These would fix carbon using photosynthesis to drive the creation of sugar, starch, and vegetable oil.

For their next level we have primary consumers, the ones that are basically vegetarians, like deer, rabbits, and squirrels. As we move up levels, think about the energy budget. If the plants convert carbon dioxide into 1000 calories of fixed carbon (ex. sugar) then the primary consumers could convert approximately 100 into their bodies (biomass). Each level is only able to use about 10% of the original biomass.

The next level is the secondary consumer or I think of it as the omnivore like the bear and human. The energy conversion is harder to generalize as time of year and amount of organisms that are consuming plants versus other animals can vary.

As you see there is a lot of wiggle room in these models. When we study trophic relationships we need to understand statistics to make sense of our models. For instance if a human decides to eat less meat or be a vegetarian then the calculation on energy saved can vary greatly. Therefore one must take large populations to study and this is where statistics comes in.

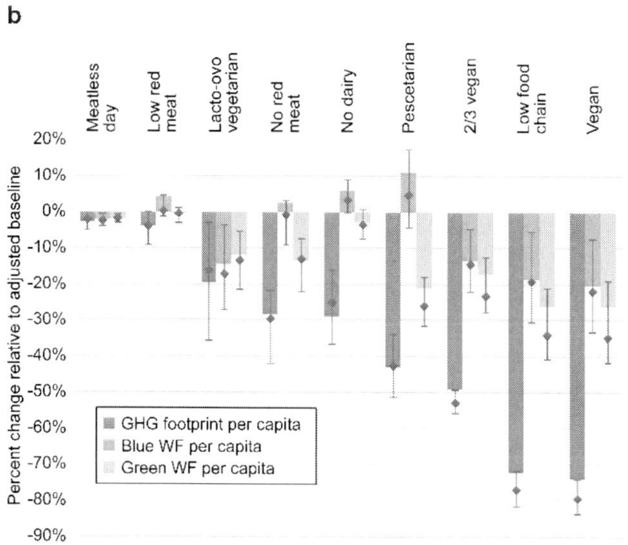

Figure 19 *Kim B.F. et al., (2020). Country-specific dietary shifts to mitigate climate and water crises.

Scientists from John Hopkins present data that compares diet modification for 140 different countries. The baseline is just your normal diet, and then other diets are compared to the normal diet to see how much less impact they have on greenhouse gas (GHG) and water. The big thing I try to teach students is how we critically analyze this data. There are lots of calculations in here to adjust and compare the diets.

First, we must understand axis. The independent axis or X axis that runs horizontally along the graph. The independent variable is what is being manipulated, or in this case the different types of diets. The dependent or Y axis or up and down axis is what is being measured based on the changing diets. Let's look specifically at greenhouse gas based on changing diet from the normal baseline. Looking at the error bars, the baseline GHG ranges between 25 and negative 10, with the mean right about zero. The error bars usually represent 95 percent of the standard deviation of the data and the mean is the average. When we look at meatless day or low red meat diets there is little difference from baseline. Next looking at lacto-ovo diets, the mean is ten percent less than the baseline but the error bars overlap the data for the baseline so that they cannot be considered statistically different. It is not until you get to Pescatarian (fish eater) that there is significantly less greenhouse gases emitted as compared to baseline. All of this goes to show that diet is complex and larger sampling size helps us understand the data, and error bars will show you how much variance there is in your data.

Getting back to understanding our simplified trophic pyramid, the last level after omnivore is the pure carnivore. By this level we are down to only 1 calorie produced from the original 1000 calories of the autotrophs. The data from the John Hopkins researchers did show that the vegan or no animal products diets used significantly less energy. This makes sense based on our understanding of the efficiency of energy conversion. But still, the data is not perfectly clean and assumptions are made just enough for me to keep eating pepperoni pizza.

On the wall of the science class I am sitting in right now there is a sign posted asking, "What refutes science?" Then it answered, "better science; not your feelings, religion, politics, or opinion after watching two videos on YouTube". This makes me think about what I say the point of science is, and that is that we always need to be learning. The complexity of our bodies and community require that we must learn.

The simplest form of learning is stimulus response or as psychologists call it classical conditioning where an involuntary response is elicited. You burn your hand on the stove and you learn not to touch the stove. The response can be controlled by hormonal or nervous system cell communication. Simply put, hot stove is nervous and hot person is hormonal (and a bit nervous).

Next is operant conditionings when you can train a response based on repetition. Dog training is always the best example as their innate behavior is probably not to come to someone shouting "come bingo".

Lastly, it is observational learning. My dad would always tell a story about me and my sister growing up that my older sister would always learn by diving in and just doing whatever. Where I, as the second child, would watch my sister and assess if it was something that I should do.

This is the behavior of the scientist to observe the world to determine cause (independent variable) and effect (dependent variable) to solve a problem.

In the end we know, by looking at the history of the world, we will constantly have a changing environment and thus biology will evolve to adapt to the environment. To have resilience in the genetic code is to have diversity. So when things change hopefully there are organisms that thrive in the new environment. When the earth cooled there were smaller hair insulated mammals to repopulate the earth after the dinosaurs passed. Or when there is a potato blight you can grow grain. Diversity is the key to survival.

Currently we are in a period of loss of diversity on our earth. We as humans, are growing exponentially and as a result we are simplifying our world to sustain our population. As we systemize agriculture and transportation, we reduce diversity. We are constantly studying how to improve supply of food and energy but as we do this we create a system dependent on high energy demand that we supply with limited fossil fuels. We have seen other systems collapse when these conditions of low diversity and expanding population of one species arise and consume the resources faster than they are replenished. It is hard to predict when the collapse will happen so we look at other systems that have gone through it to learn by observation.

Like the dust bowl in the Midwest, United States in the 1930s where people had converted the drought tolerant grasslands to crop land. When there were a few years of drought, the crops did not grow and left the soil bare, so the wind just blew the soil away. After the agricultural boom, there was a problem as soil was depleted and agriculture suffered.

Could the dust bowl have been predicted? The water cycle is somewhat predictable. We know that winter will be the coldest and summer the warmest, so depending on where you are you probably have predicted wet times of year. Meteorologists work to predict the weather and they are quite good at predicting the near future but as they try and predict further into the future the model accuracy degrades. I used to look at the farmer's almanac that predicted the seasonal variation. The National Oceanic and Atmospheric Administration (NOAA) predicts both weather (short term) and climate (long term) by tracking and measuring water around the globe. Water is not the only cycle we track. Carbon, nitrogen, phosphorus, and sulfur are some of the element cycles we track as these are the elements of life. By tracking these elements we can predict the cause or effect of a change to one or more components of the biological system.

The carbon cycle is the one people are most familiar with other than the water cycle. News these days covers the effect carbon dioxide has on the temperature of our earth. Carbon cycle is basically the conversion between fixed organic carbon in biological form to inorganic carbon of carbon dioxide or carbon monoxide in the atmosphere.

Understanding these cycles helps us know what we need to do to manipulate a system or plan for future changes. All biological systems work with these elements that we can track. The unity of biology is amazing. From element, to macromolecule, to cell, to tissue, to organ system, to organism, to ecosystem, to biosphere, the biological stuff is amazing. Change is explained by natural processes that can be predicted and manipulated. We are all linked by the primordial life we evolved from. This is all B. S. We just need to learn to listen to the communications that bring us harmony and balance.

What are the unifying themes of biology? How can we predict the cause and effect of biological systems? How do we study a system and how do we know if the data is significant or not? How do we make decisions based on science?

The End.

About the Author

Gretchen Watkins is a scientist and educator with degrees in biology and environmental engineering. She is an award-winning science high school teacher, with a preceding career in the field of environmental science. She has worked for high schools, universities, state, tribal and federal governments, and written laws to protect human health and the environment. In *It's All BS*, Gretchen brings you real-life stories from her work in the field, that make it easy to understand biology, and how it is central to all human life. Gretchen wants you to know, the better you understand life, the better you live life.

Acknowledgements

I owe a tremendous amount of gratitude to many individuals who helped me at various stages of putting this book together. First to my husband who was my biggest supporter and inspiration. Also, my father (Jack Wilson) and father-in-law (Clyde Watkins), for their editorial assistance. Lastly, all my science friends and students who drive me to understand life better. Thank you!

Figure 19: *Kim, B. F., Santo, R. E., Scatterday, A. P., Fry, J. P., Synk, C. M., Cebron, S. R., ... & Nachman, K. E. (2020). Country-specific dietary shifts to mitigate climate and water crises. Global environmental change, 62, 101926.

Made in United States
North Haven, CT
04 November 2026

21772151R00062